Lecture Notes in Nanoscale Science and Technology

Lecture Notes in Nanoscale Science and Technology (LNNST) aims to report latest developments in nanoscale science and technology research and teaching – quickly, informally and at a high level. Through publication, LNNST commits to serve the open communication of scientific and technological advances in the creation and use of objects at the nanometer scale, crossing the boundaries of physics, materials science, biology, chemistry, and engineering. Certainly, while historically the mysteries in each of the sciences have been very different, they have all required a relentless step-by-step pursuit to uncover the answer to a challenging scientific question, but recently many of the answers have brought questions that lie at the boundaries between the life sciences and the physical sciences and between what is fundamental and what is application. This is no accident since recent research in the physical and life sciences have each independently cut a path to the edge of their disciplines. As both paths intersect one may ask if transport of material in a cell is biology or is it physics? This intersection of curiosity makes us realize that nanoscience and technology crosses many if not all disciplines. It is this market that the proposed series of lecture notes targets.

Shama Prasada Kabekkodu
Rajib Biswas • Nirmal Mazumder
Aniruddha Chatterjee
Editors

Functional Quantum Dots

Synthesis, Modification, and Biomedical Applications

Springer

Editors
Shama Prasada Kabekkodu
Manipal School of Life Sciences
Manipal Academy of Higher Education
Manipal, Karnataka, India

Rajib Biswas
Department of Physics
Tezpur University
Tezpur, Assam, India

Nirmal Mazumder
Department of Biophysics
Manipal School of Life Sciences
Manipal Academy of Higher Education
Manipal, Karnataka, India

Aniruddha Chatterjee
Dunedin School of Medicine
University of Otago
Dunedin, Otago, New Zealand

ISSN 2195-2159 ISSN 2195-2167 (electronic)
Lecture Notes in Nanoscale Science and Technology
ISBN 978-3-031-98218-7 ISBN 978-3-031-98219-4 (eBook)
https://doi.org/10.1007/978-3-031-98219-4

This Springer imprint is published by the registered company Springer Nature Switzerland AG
The registered company address is: Gewerbestrasse 11, 6330 Cham, Switzerland

Preface

Quantum dots are semiconductor nanocrystals that exhibit remarkable optical and electronic properties due to quantum confinement effects arising from their nanoscale dimensions (typically 1–10 nm). These unique characteristics have made them a focal point of interdisciplinary research in physics, chemistry, materials science, and engineering.

This book comprehensively explores the synthesis, modification, and biological applications of quantum dots across ten chapters authored by field experts. Key topics include:

- Structural and optoelectronic properties of quantum dots
- Applications in optical imaging, biomedicine, and cancer diagnosis
- Toxicity mitigation strategies
- Photodynamic therapy
- Sensing technologies using perovskite and boron nitride quantum dots

Chapter Highlights:

1. **Roy et al.** detail the fundamental structural and optoelectronic properties of quantum dots, emphasizing excitonic behavior and energy-level quantization.
2. **Phadke et al.** discuss the transformative role of quantum dots in optical imaging for biomedical research.
3. **Nath et al.** review the historical evolution, synthesis methods, and characterization techniques of quantum dots for biomedical applications, including future challenges.
4. **Sridhar et al.** address the functional quantum dots' biomedical uses (imaging, drug delivery, and biosensing) and toxicity mitigation through surface modifications.
5. **Balaji et al.** present a quantum dot-enhanced localized surface plasmon resonance for biomedical imaging and therapy.
6. **Ajjanagadde et al.** focus on cadmium telluride quantum dots in cancer diagnosis, gamma-ray detection, and bioimaging.

7. **Dey et al.** explore the quantum dot-based optical microscopy for oral cancer detection.
8. **Shetty et al.** examine quantum dots in photodynamic therapy, covering synthesis and in vitro/vivo applications.
9. **Saikia et al.** highlight the perovskite quantum dots as sensor probes for target analytes.
10. **Gogoi et al.** summarize the advances in boron nitride quantum dots for biomedicine.

Manipal, Karnataka, India — Shama Prasada Kabekkodu
Tezpur, Assam, India — Rajib Biswas
Manipal, Karnataka, India — Nirmal Mazumder
Dunedin, Otago, New Zealand — Aniruddha Chatterjee

Acknowledgments

We thank the authors for their valuable contribution to this book. Special thanks to Ms. Tina Silva and Ms. Nemul Khan from Springer Nature for extending all the support and timely help in getting the book published. We hope that this book will stimulate the readers and their interest in the quantum dots and optical imaging.

Contents

Editors and Contributors

About the Editors

Shama Prasada Kabekkodu is a Professor at the Department of Cell and Molecular Biology, Manipal School of Life Sciences, Manipal Academy of Higher Education (MAHE), Manipal, India. He obtained his Ph.D. in 2011 from MAHE. He has 14 years of experience in teaching and research. He has published 136 articles and 7 conference proceedings in peer-reviewed international journals. He has published 3 patents, edited 1 book, 11 book chapters, and 2 general articles. His research interests include functional epigenetics, cancer biology, and signaling. He has received research funding from various Indian funding agencies, including Wellcome Trust DBT India Alliance Intermediate Fellowship in 2023. He is a Life Member of the Society of Biological Chemists, India, and the Society for Mitochondrial Research and Medicine, India.

Rajib Biswas is an Associate Professor in the Department of Physics, School of Sciences, Tezpur University, India. He pursued his Ph.D. from CSIR-NEIST, India. He has 15 years of teaching experience and 17 years of research experience. He has published 150 articles in peer-reviewed international journals along with 23 book chapters. He has also published three patents and has edited six books. His research interests include applied photonics, quantum dots, nanomaterials, and material photonics. He has received research funding from various funding agencies, including BIRAC and MoES. He is a Life Member of the Earthquake Society of India, Indian Physics Association and an emeritus member of the Optical Society of America. Currently, he serves as an associate editor for the photonics category in *Frontiers in Physics* and holds editorial roles in many peer-reviewed journals.

Nirmal Mazumder is an Associate Professor at the Department of Biophysics, Manipal School of Life Sciences, Manipal Academy of Higher Education (MAHE), Manipal, India. He obtained his Ph.D. in 2013 from the Institute of Biophotonics, National Yang Ming Chiao Tung University, Taipei, Taiwan, under the guidance of

Prof. Fu-Jen Kao. From 2013 to 2016, he worked as a postdoctoral fellow at the University of Virginia, USA, and the Italian Institute of Technology, Genoa, Italy. He has been developing nonlinear optical microscopes, including two photon fluorescence, second-harmonic generation, and coherent anti-Stokes Raman scattering for biomedical applications. His research interests include the development of microscopy–spectroscopy techniques, biomedical devices, biophotonics, smart materials, microfluidics, and its biomedical applications. He has published more than 125 research articles in peer-reviewed international journals, 16 books, 45 book chapters, and several poster, oral, and invited presentations at national and international conferences. He has been recognized as a top scientist, ranking in the top 2% by c-score in the sub-field for the years 2023, 2024 and 2025. He is working on several national and bilateral research projects funded by the Government of India as a Principal Investigator. He is also working as a reviewer, guest editor, and associate editor in several prestigious journals. He is a member of several national and international scientific societies and organizations, including the Optical Society (OPTICA) (Senior member), SPIE—the International Society for Optical Engineering (Senior Member), the Society of Biological Chemists (I) (Life Member), the Environmental Mutagen Society of India (Life Member), the Asian Polymer Association (Life Member), the Indian Laser Association-ILA (Life Member), and the Research Society for the Study of Diabetes in India (RSSDI, Life Member).

Aniruddha Chatterjee is a Research Associate Professor, Department of Pathology, Dunedin School of Medicine, University of Otago, New Zealand. He obtained his Ph.D. from the University of Otago in 2013. His research focuses on epigenetics, molecular biology, and the bioinformatic analysis of next-generation sequencing data. He is the recipient of several awards and honors, including the Roche Translational Cancer Research Fellowship from the New Zealand Society of Oncology, Wellcome Trust (UK) scholarship for an advanced course in genomics, a visiting biomedical fellowship from the NZ-China Centre, and visiting fellowships from the JSPS, Japan, and the Royal Society of New Zealand. He has published 89 articles in peer-reviewed journal, authored a book, 10 conference proceedings, and delivered 53 invited talks.

Contributors

Asfi Ahmed Department of Chemical Sciences, Tezpur University, Napaam, Tezpur, Assam, India

Gazi Ameen Ahmed Laboratory for Plasma Processing of Materials, Department of Physics, Tezpur University, Tezpur, India

Anagha Ajjanagadde Department of Cell and Molecular Biology, Manipal School of Life Sciences, Manipal Academy of Higher Education, Manipal, Karnataka, India

Medha Bairy Department of Cell and Molecular Biology, Manipal School of Life Sciences, Manipal Academy of Higher Education, Manipal, Karnataka, India

Roopal Behera Department of Cell and Molecular Biology, Manipal School of Life Sciences, Manipal Academy of Higher Education, Manipal, Karnataka, India

Pronama Biswas School of Basic and Applied Sciences, Dayananda Sagar University, Bengaluru, India

Rajib Biswas Laboratory for Plasma Processing of Materials, Department of Physics, Tezpur University, Tezpur, India

Department of Physics, Tezpur University, Tezpur, Assam, India

Manash R. Das Materials Sciences and Technology Division, CSIR-North East Institute of Science and Technology, Jorhat, Assam, India

Shyamal K. Das Department of Physics, Tezpur University, Tezpur, Assam, India

Belaguppa Manjunath Ashwin Desai School of Engineering, Dayananda Sagar University, Bengaluru, India

Doyal Dey Department of Public Health Genomics, Manipal School of Life Sciences, Manipal Academy of Higher Education, Manipal, Karnataka, India

Swapan K. Dolui Department of Chemical Sciences, Tezpur University, Napaam, Tezpur, Assam, India

Suvarna Ghosh School of Basic and Applied Sciences, Dayananda Sagar University, Bengaluru, India

Debika Gogoi Materials Sciences and Technology Division, CSIR-North East Institute of Science and Technology, Jorhat, Assam, India

Devipriya Gogoi Materials Sciences and Technology Division, CSIR-North East Institute of Science and Technology, Jorhat, Assam, India

Mrudula Gore Department of Cell and Molecular Biology, Manipal School of Life Sciences, Manipal Academy of Higher Education, Manipal, Karnataka, India

Rajat Hegde Department of Cell and Molecular Biology, Manipal School of Life Sciences, Manipal Academy of Higher Education, Manipal, Karnataka, India

Shama Prasada Kabekkodu Department of Cell and Molecular Biology, Manipal School of Life Sciences, Manipal Academy of Higher Education, Manipal, Karnataka, India

Arushi Kapoor School of Civil & Environmental Engineering and Construction Management, University of Texas at San Antonio, San Antonio, TX, USA

Ayush Khandelwal Department of Cell and Molecular Biology, Manipal School of Life Sciences, Manipal Academy of Higher Education, Manipal, Karnataka, India

Sanjeev Pran Mahanta Department of Chemical Sciences, Tezpur University, Napaam, Tezpur, Assam, India

Centre for Multidisciplinary Research, Tezpur University, Napaam, Tezpur, Assam, India

Nirmal Mazumder Department of Biophysics, Manipal School of Life Sciences, Manipal Academy of Higher Education, Manipal, Karnataka, India

Kaushik Kokil Nath Laboratory for Plasma Processing of Materials, Department of Physics, Tezpur University, Tezpur, India

K. Sreedhara Ranganath Pai Department of Pharmacology, Manipal College of Pharmaceutical Sciences, Manipal Academy of Higher Education, Manipal, Karnataka, India

Ajeetkumar Patil Manipal Institute of Applied Physics, Manipal Academy of Higher Education, Manipal, India

Varad Mahesh Phadke Manipal Institute of Applied Physics, Manipal Academy of Higher Education, Manipal, India

Alevoor Srinivas Bharath Prasad Department of Public Health Genomics, Manipal School of Life Sciences, Manipal Academy of Higher Education, Manipal, Karnataka, India

Konica Roy Department of Physics, Tezpur University, Tezpur, Assam, India

Priyankamoni Saikia Department of Chemical Sciences, Tezpur University, Napaam, Tezpur, Assam, India

Aishath Shaheeda Department of Cell and Molecular Biology, Manipal School of Life Sciences, Manipal Academy of Higher Education, Manipal, Karnataka, India

Sachin Shetty Department of Cell and Molecular Biology, Manipal School of Life Sciences, Manipal Academy of Higher Education, Manipal, Karnataka, India

Nishant Shukla Department of Physics, The Assam Kaziranga University, Jorhat, Assam, India

Bhoomika Sridhar School of Basic and Applied Sciences, Dayananda Sagar University, Bengaluru, India

Merla Sudha School of Basic and Applied Sciences, Dayananda Sagar University, Bengaluru, India

E. Vignesh Balaji Department of Pharmacology, Manipal College of Pharmaceutical Sciences, Manipal Academy of Higher Education, Manipal, Karnataka, India

Structural and Optoelectronic Properties of the Quantum Dots

Konica Roy, Nishant Shukla, and Shyamal K. Das

1 Introduction

In recent years, nanotechnology has made significant strides across diverse fields, such as medicine, chemistry, physics, and engineering, driven largely by advancements in quantum dots (QDs) [1–6]. QDs, often dubbed “artificial atoms” owing to their unique optoelectronic properties, have a history dating back to 1981, when Alexey Ekimov first identified them within a glass matrix [7]. This discovery marked the beginning of a transformative journey in which Louis Brus synthesized the first solution of a colloidal nanocrystalline semiconductor in 1985 [8]. Later, in 1988, Mark Arthur Reed coined the term “quantum dots” and demonstrated their photoluminescent capabilities [9]. These nanoscale semiconductor particles, which are smaller than 10 nm, exhibit size-dependent optical and electronic characteristics that are fundamentally distinct from those of bulk materials [10, 11]. Quantum confinement confines electrons and holes within QDs to regions comparable to their de Broglie wavelength, resulting in discrete, quantized energy states. This confinement gives quantum dots exceptional properties, including high brightness, quantum yield, photostability, and tunable excitation and emission spectra, outperforming traditional devices [12, 13]. Historically, ancient civilizations unwittingly utilized materials akin to QDs. The Romans and Greeks employed substances such as PbO and $Ca(OH)_2$ in cosmetics, inadvertently harnessing early forms of nanomaterials. In the early twentieth century, CdS and CdSe nanoparticles were rapidly introduced into silicate glasses, revolutionizing the control of glass color through nanotechnology [14, 15]. The pivotal moment occurred in 1981 when Ekimov and Onushchenko

K. Roy (✉) · S. K. Das
Department of Physics, Tezpur University, Tezpur, Assam, India

N. Shukla
Department of Physics, The Assam Kaziranga University, Jorhat, Assam, India

S. P. Kabekkodu et al. (eds.), *Functional Quantum Dots*, Lecture Notes in Nanoscale Science and Technology 40,
https://doi.org/10.1007/978-3-031-98219-4_1

observed a blueshift in the optical spectrum of CuCl within silicate glass, highlighting the effects of quantum confinement [16]. This breakthrough laid the groundwork for modern synthesis techniques, enabling precise control over QD size and composition. Efros and Efros's 1982 hypothesis on harnessing quantum size effects to manipulate the optical properties of glass further propelled QD research [17], whereas Rosetti et al.'s 1991 work emphasized the size-dependent attributes of colloidal semiconductor solutions [18]. Since Ekimov's pioneering work on CdSe in the 1980s, researchers have expanded into CdTe, InAs, and InP, each offering distinct bandgap properties suitable for various applications [19–22]. Nonalcadmium alternatives such as $CuInS_2$ and ZnS, which are known for their relatively low toxicity, have gained prominence in biomedical uses [23–25]. Ag-In-S QDs excelled in terms of light absorption and emission with prolonged photoluminescence lifetimes and high quantum yields [26]. Newer ternary I-III-VI QDs, such as copper indium sulfide (CIS), provide adaptable bandgap adjustments crucial for solar cells and LEDs, coupled with stability and ease of synthesis beneficial for biophotonics [27, 28]. The incorporation of QDs into matrices enhances their stability, mechanical properties, and application versatility. For example, compared with standalone CdZnTe QDs, CdZnTe QDs in a $CaCO_3$ matrix exhibited superior photostability and processability [29]. Perovskite QDs (PQDs) have emerged as promising semiconductor nanoparticles with high quantum yields, narrow emission spectra, and superior color purity and are poised to revolutionize optoelectronics [6, 30]. The use of tin oxide QDs as anode materials allows high charge capacity in lithium-ion batteries [31].

Experimental and theoretical strides over the last two decades have delved deeply into its fundamental properties, driven by its potential across various industries. Synthesis methods have played pivotal roles, facilitating the creation of highly crystalline Si and III-V semiconductors (e.g., GaAs, InP) [32–34]. The tunability of these systems, governed by the QD size, enhances the signal brightness, quantum yield, and photostability, which is ideal for applications in fields such as imaging and sensing [35–37]. Moreover, QDs offer unique electronic properties; unlike bulk semiconductors, their tunability via quantum confinement induces a blueshift in the absorption and emission spectra with decreasing size, enabling tailored optoelectronic device designs [38]. The versatility of QDs has also fostered their broad adoption across fields such as medicine, where they promise breakthroughs in diagnostic imaging and disease detection through bright, stable fluorescence [39–41]. In sensing, QDs facilitate sensitive, selective detection via fluorescence-based sensors. In solar cells, QDs bolster light absorption efficiency, advancing renewable energy technologies [42–45]. In electrochemistry, they demonstrate positive impacts on critical performance factors such as discharge capacitance, the reaction process, ion diffusion kinetics and long-term cycling stability. Owing to their beneficial impact on electrolytes used in energy storage systems, carbon- and graphene-based quantum dots are promising materials. The various functional groups of these materials enable them to serve effectively as electrolyte additives. In composite systems, the structure of these QDs shortens the ion diffusion path, providing more active sites for host ions. This structure improves the fast transport of electrons and helps prevent the dissolution of discharged products [46, 47]. QDs encompass a range of materials tailored to specific applications. Quantum dots (QDs) are at the forefront of nanotechnology, offering unique quantum confinement effects and diverse applications across various fields. From their initial discovery to recent innovations, QDs continue to transform medicine, electronics, and more, owing to their customizable properties that unlock new scientific and technological possibilities. While the properties of single electrons are intriguing, most mesoscopic systems naturally contain multiple electrons unless they are specifically engineered to confine just one electron at a time, as in the case of the electron turnstile. Collective effects play a major role in shaping the optoelectronic properties of these systems, especially where the electron mean free path of the electrons exceeds the device size (ballistic regime). Understanding electron-correlation effects is essential for both deepening our fundamental knowledge and developing future practical applications. The expectation that reducing size will lead to faster and more heat-efficient electronic and optoelectronic devices than current technologies primarily motivates researchers to study mesoscopic systems. This chapter focuses on understanding the underlying physics of these systems through mathematical models and various QD applications.

2 The Quantum Phenomenon in Low-Dimensional Structures

Nanoscale science examines materials with dimensions between 1 and 100 nm, which are classified as low-dimensional structures. Dimensionality describes how many directions carriers can move freely within a material. In a 3D structure, carriers can move freely in all directions, as in bulk materials. When a bulk material is restricted in one dimension, it forms a thin film, creating a two-dimensional (2D) structure or quantum well, where carriers are confined in one direction. If confinement is applied in an additional direction, the material transitions into a one-dimensional (1D) structure, or quantum wire, where carriers are restricted to movement in a single line. Further confinement in all three dimensions results in a zero-dimensional (0D) structure, commonly referred to as a quantum dot, where carriers are confined in all spatial directions. This progression through low-dimensional structures is shown in Fig. 1. The significance of spatial confinement for electronic carriers becomes evident when the density of states in various low-dimensional structures is examined. The figure illustrates how the density of states shifts as a material transitions from an unconstrained bulk form to fully confined three-dimensional states. Confinement leads to substantial alterations in the allowed energy levels for carriers. In bulk materials, carriers occupy nearly continuous energy bands, whereas in a 0D quantum dot, carriers are restricted to discrete, fully quantized energy levels. As shown in Table 1, the confinement effect in quantum dot systems can be explored by examining different geometrical configurations and their impact on the energy levels of electrons. Starting with a general quantum wire,

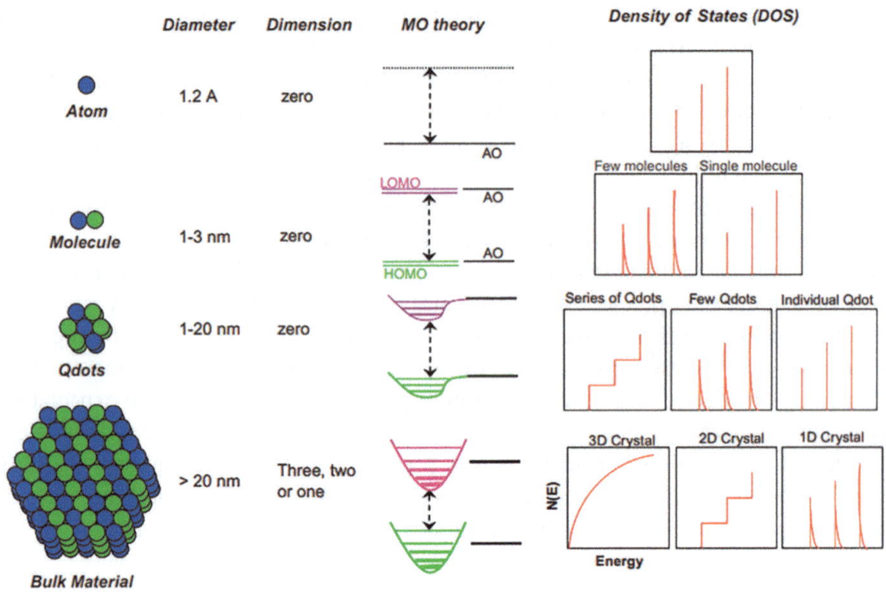

Fig. 1 Schematic diagram depicting the variation in the density of quantum states (DOS) as the number of atoms in a material changes. AO stands for the atomic orbital, and MO stands for the molecular orbital. Reproduced with permission, copyright 2008, Wiley-VCH [48]

one can understand how confinement influences energy levels before delving into specific cases, such as wires with rectangular cross-sections and spherical quantum dots. In a quantum wire, the three-dimensional Schrodinger equation with a constant effective mass becomes simplified when the motion along the wire's length (x-axis) is separated from the motion in transverse directions.

The total potential $V(x,y,z)$ can be separated into a one-dimensional component along the length of the wire and a two-dimensional confinement potential that acts perpendicular to the wire. This separation leads to two separate equations: one represents plane wave propagation along the wire, whereas the other represents the Schrodinger equation for confinement in the two-dimensional plane perpendicular to the wire. For a quantum wire with a rectangular cross-section surrounded by infinite barriers, the potential is zero inside the wire and infinite outside, causing the wave function to be zero at the boundaries. This setup means that the Schrodinger equation is applicable only within the wire, with confinement characterized by two principal quantum numbers n_y and n_z. The energy levels $E_{y,z}$ decrease as the wire size increases. In a wire with a square cross-section, the energy levels for the (1, 2) and (2, 1) states are degenerate. However, in a wire with a rectangular cross-section, where $L_y \neq L_z$, this degeneracy is no longer observed. Considering more practical scenarios, a rectangular quantum wire is then considered with finite barriers. Here, the potential $V(y,z)$ cannot be separated into independent components $V(y)$ and $V(z)$. However, the wave functions and confinement energies are analogous to those found in a finite quantum well. The energy curve in this case shows similar behavior to that

Table 1 The governing equations for various quantum confined structures

Schrödinger's equation

The general three-dimensional Schrödinger equation for constant effective mass is:

$$\frac{-\hbar^2}{2m^*}\nabla^2\psi(x,y,z)+V(x,y,z)\psi(x,y,z)=E\psi(x,y,z)$$

$$\frac{-\hbar^2}{2m^*}\frac{\partial^2\psi(x)}{\partial x^2}=E_x\psi(x);\quad E_x=\frac{\hbar^2k_x^2}{2m^*}$$

$$\frac{-\hbar^2}{2m^*}\left[\frac{\partial^2\psi(y,z)}{\partial y^2}+\frac{\partial^2\psi(y,z)}{\partial z^2}\right]+V(y,z)\psi(y,z)=E_{y,z}\psi(y,z)$$

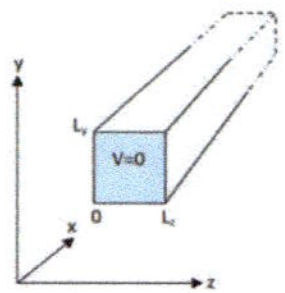 *Infinite rectangular wires*	$\frac{-\hbar^2}{2m^*}\left[\frac{\partial^2\psi(y,z)}{\partial y^2}+\frac{\partial^2\psi(y,z)}{\partial z^2}\right]+V(y,z)\psi(y,z)=E_{y,z}\psi(y,z)$ $\psi(y,z)=\psi(y)\psi(z):\ E_{y,z}=\frac{\hbar^2\pi^2}{2m^*}\left[\frac{n_y^2}{L_y^2}+\frac{n_z^2}{L_z^2}\right]$
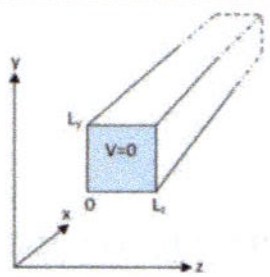 Finite rectangular wire	$\frac{-\hbar^2}{2m^*}\left[\psi(z)\frac{\partial^2\psi(y)}{\partial y^2}+\psi(y)\frac{\partial^2\psi(z)}{\partial z^2}\right]+\left[V(y)+V(z)\right]\psi(y)\psi(z)=E_{y,z}\psi(y)\psi(z)$ $\frac{-\hbar^2}{2m^*}\frac{\partial^2\psi(y)}{\partial y^2}+V(y)\psi(y)=E_y\psi(y)$ $\frac{-\hbar^2}{2m^*}\frac{\partial^2\psi(z)}{\partial z^2}+V(z)\psi(z)=E_z\psi(z)$

(Continued)

Table 1 (Continued)

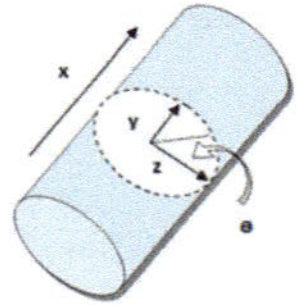 Circular cross-section wire	$$\frac{-\hbar^2}{2m^*}\left[\frac{\partial^2\psi(y,z)}{\partial y^2}+\frac{\partial^2\psi(y,z)}{\partial z^2}\right]+V(y,z)\psi(y,z)=E_{y,z}\psi(y,z)$$ In spherical coordinates, $$\frac{-\hbar^2}{2m^*}\left[\frac{1}{r}\frac{\partial}{\partial r}+\frac{\partial^2}{\partial r^2}\right]\psi(r)+V(r)\psi(r)=E_r\psi(r)$$
Spherical quantum dots 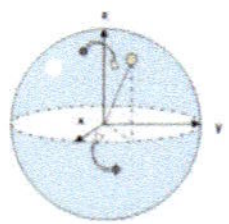	$$\frac{-\hbar^2}{2m^*}\left[\frac{2}{r}\frac{\partial}{\partial r}+\frac{\partial^2}{\partial r^2}\right]\psi(r)+V(r)\psi(r)=E_r\psi(r)$$

in the infinite-barrier scenario, with energy levels exhibiting degeneracy between the (1, 2) and (2, 1) states and the eigenvalue decreasing as the wire side length increases.

The energy curve in this case shows similar behavior to that in the infinite-barrier scenario, with energy levels exhibiting degeneracy between the (1, 2) and (2, 1) states and the eigenvalue decreasing as the wire side length increases. Examining the Schrodinger equation for a quantum wire's confined cross-section shows that the exact form of the kinetic energy operator is essential, assuming a constant effective mass. The energy curve shows that as the wire radius increases, the confinement energy decreases, with odd-parity eigenstates having higher energy than even-parity eigenstates. Likewise, a spherical quantum dot, which is easier to manage than a cuboid symmetry finite-barrier dot, is analyzed similarly to a quantum wire with a circular cross-section. For spherical quantum dots, the results from the three lowest energy levels confirm that, as anticipated, the confinement energy decreases as the size of the dot increases, aligning with the behavior observed in other confined systems.

As one moves to lower confined states, the potential becomes increasingly significant in determining the energy levels. In semiconductor materials, the bandgap energy defines the minimum amount of energy needed to create free carriers when they are optically excited. However, excitations occurring just below this energy threshold have also been observed, as in low-temperature absorption studies. This is attributed to the formation of excitons, which are bound states of an electron and a hole. The energy necessary to generate an exciton is lower than the bandgap energy because, in an exciton, the electron and hole are bound together in a stable state. In this configuration, the electron orbits around the hole, which creates a potential well in which the electron is confined. This binding reduces the overall energy required compared with the energy needed to produce free carriers across the bandgap. In spatially confined systems such as quantum dots, excitons play a crucial role in forming quantized energy states. These excitons behave similarly to hydrogen atoms, where Coulombic forces attract the electron to the nucleus. Similarly, in a quantum dot, the electron and hole form a hydrogen-like system, with the electron orbiting the hole. The exciton Bohr diameter defines the size of the exciton and serves as a key parameter for assessing quantum confinement effects within the material. As the quantum dot size approaches this diameter, confinement effects become significant. The quantum dots exhibit different confinement regimes on the basis of their size relative to the exciton Bohr diameter. Under weak confinement, which occurs when the dot size is approximately 3–10 times the exciton Bohr radius, the Coulomb interaction between the electron and hole remains significant. In this context, the electron and hole are considered a correlated pair, with their center-of-mass motion being quantized owing to the confinement potential. In this regime, the energy levels can be calculated via the parabolic approximation of the Schrodinger equation, which assumes an infinite potential barrier at the boundary of the quantum dot. As the quantum dot shrinks and its size approaches the exciton Bohr radius, the confinement effects shift from being weak to strong. When the quantum dot size is smaller than the exciton Bohr radius, strong confinement occurs.

In this regime, the quantized kinetic energy of the carriers overshadows the Coulomb interactions between the electron and hole. The electron and hole wave functions become uncorrelated, and their energy states are described via independent Bessel wave functions. Despite the changes in confinement, a consistent relationship between energy and dot size results in a blueshift as the dot size decreases. The energy expressions for strongly confined systems and weakly confined systems differ owing to the presence of Bessel functions and variations in the effective carrier mass, which impact the quantization of energy levels. These differences are material dependent, reflecting variations in electron and hole masses across different semiconductors. Overall, understanding these confinement effects is essential for optimizing quantum dot-based devices, as they directly influence the electronic and optical properties of the material.

3 Quantum Dot Device Applications

Research on semiconductor quantum dot systems has revealed fascinating alterations in the behavior of these zero-dimensional structures. Numerous concepts have been proposed to leverage the unique properties of quantum dots for specialized applications. The integration of quantum wells into device architectures has already led to substantial improvements in the performance of various applications. The incorporation or replacement of components with quantum dots in devices is expected to markedly increase their performance beyond current standards. The study of quantum dots covers a wide range of areas, including chemical gas sensing, optoelectronics, nonlinear optics, telecommunications, and computing. This section highlights three of the most promising applications: QDs for energy storage applications, QD materials for optoelectronic applications and QD materials for biomedical applications.

3.1 QDs for Energy Storage Applications

The choice of electrode materials and their compatibility with the selected electrolyte largely determine the performance and efficiency of electrochemical energy storage systems. Recent advancements in battery technology have been made through the incorporation of QD-based electrode materials, which have shown significant potential in enhancing energy storage systems [49–53] (Fig. 2). It offers numerous advantages due to its abundance of heteroatomic functional groups, which provide a variety of active sites. These properties make QDs excellent candidates for use as composite materials in both current collectors and active electrodes. By enhancing the ionic conductivity, enabling high charge and discharge rates, providing substantial capacity, and ensuring stable cycling performance, QDs significantly enhance the overall functionality and reliability of electrochemical energy

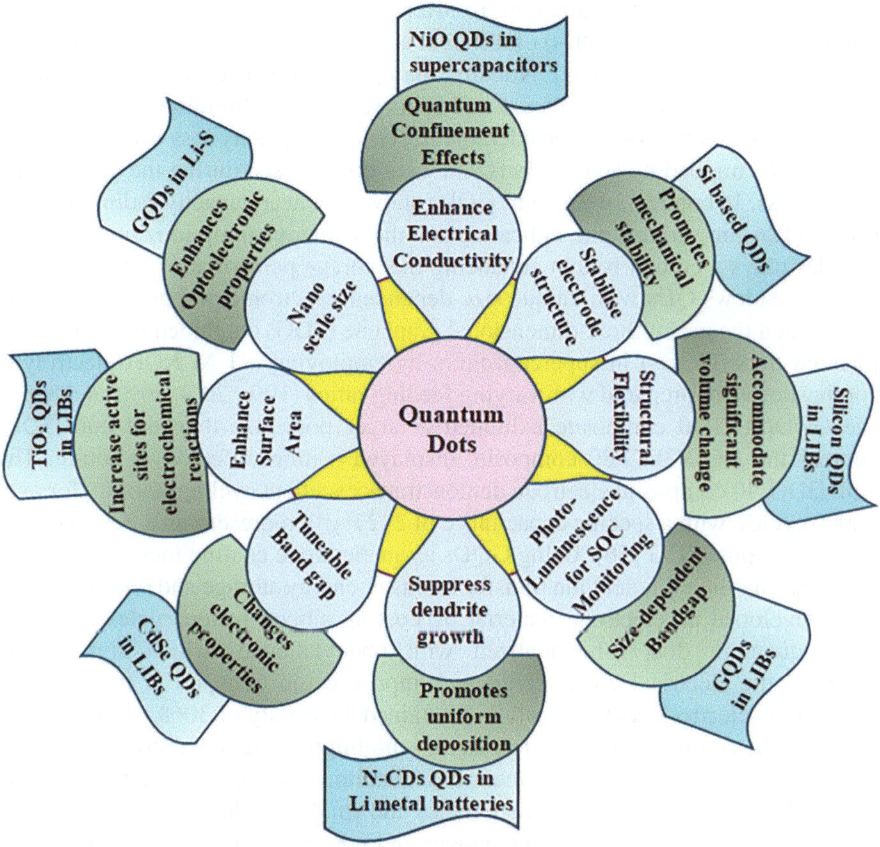

Fig. 2 Schematic illustration of the characteristics and applications of quantum dots in energy storage devices. SOC stands for the state of charge, NiO stands for nickel oxide, N-CDs stand for nitrogen-doped carbon dots, and CdSe stands for cadmium selenide

storage devices. Battery technology incorporates QD-based electrode materials in lithium-oxygen ($Li-O_2$) batteries, which are highly recognized systems with exceptional energy densities of approximately 3 kWh kg^{-1}. To address challenges such as long-term cycling stability and high energy loss, Veeramani et al. explored a composite prepared with selenide and zinc sulfide quantum dots combined with carbon nanotubes. Recently, the issue of overpotential in $Li-O_2$ batteries has been explored using cadmium selenide and zinc sulfide QDs combined with carbon nanotubes for $Li-O_2$ batteries. The narrow bandgap semiconductor material modifies the optical properties, leading to improved capacity. As a result, the composite of QDs and CNTs increases the efficiency of electron–hole pair conversion and provides pathways for the transport of lithium ions and oxygen gas. The TEM image confirms the defect-free composite, and this combination improves the overpotential, which results in a charging potential of 2.65–4 V and cyclic stability for more than 100 cycles [54]. The size of quantum dot electrode materials greatly affects the efficiency of ion storage. Kundu et al. investigated the Li-ion storage capacity of graphene quantum dots (GQDs) of different sizes (4, 12, and 26 nm) for application in lithium-ion batteries (LIBs). Notably, the optical properties and dimensions of GQDs change with the intercalation and deintercalation of lithium ions. In particular, the 4 nm GQDs showed considerable irreversible capacity loss because lithium ions became trapped by edge defects and were not released during the deintercalation process. In contrast, the 12-nm GQDs showed slower capacity fading than the 4-nm and 26-nm variants did, indicating that the technique of optimizing the size of the QDs plays a crucial role in enhancing the storage performance [55]. The confinement endows QDs with unique size-dependent electronic properties. Luo et al. investigated the use of three-dimensional graphene (3DG) combined with graphene quantum dots (GQDs) in supercapacitors by employing a 1 M KOH electrolyte. Composites were prepared with varying feeding ratios (10%, 20%, 40%, and 80%). The GQD/3DG-40 composite exhibited a larger pore size than did pure 3DG, whereas the GQD/3DG-80 composite displayed a more effective structure. The GQD/3DG-40 composite electrode demonstrates excellent cycling stability for over 10,000 cycles, with a specific capacitance of 242 F g^{-1}, representing a 22% increase over that of pure 3DG [56]. Using GQDs as an electrode coating increases the surface area, enabling efficient ion transfer for faster energy storage and release. Lijuan et al. developed an electrode material by coating silicon nanoparticles with graphene quantum dots functionalized with phenylalanine, which significantly enhanced the electrical conductivity and improved ion transport at the interface between the electrode and electrolyte. It retained a capacity of 3068 mA h g^{-1} after 100 cycles at a current density of 100 mA g^{-1}, highlighting its suitability for lithium-ion battery applications [57]. Furthermore, quantum dots and their hybrids, owing to their distinctive electronic characteristics and small size, have been successfully utilized in stabilizing lithium metal anodes and preventing dendrite formation. For lithium–sulfur (Li–S) batteries, lithium metal is considered crucial for achieving long-lasting, high-performance systems. In this system, the formation of lithium dendrites poses risks such as short circuits and safety hazards. They also contribute to thick solid electrolyte interphase (SEI) films and side reactions, which degrade battery efficiency. Li et al. developed nitrogen and sulfur codoped carbon quantum dots (CQDs) as effective electrolyte additives for Li-S batteries to promote uniform lithium-ion deposition. These CQDs absorb Li^+ ions from the surrounding electrolyte and ensure even deposition on the current collector. Unlike untreated electrolytes that lead to random lithium-ion deposition, the nitrogen and sulfur atoms in the CQDs provide strong adsorption sites for lithium ions, enabling consistent and uniform deposition. The inclusion of this additive greatly enhanced the cycling stability of the symmetric cell, allowing it to sustain performance for more than 1200 hours [58]. Recently, Daryabari et al. conducted a theoretical study on four different quantum dots as potential anode materials for sodium-ion batteries. They concluded that fluorine-doped defected graphene quantum dots (dGQDs) hold considerable potential because of their enhanced charge distribution at edge atoms and high electrical conductivity. As the energy gap (Eg) narrows, the electrical conductivity of these F-doped dGQDs improves, which in turn reduces heat generation and enhances current flow. The strong HOMO–LUMO interactions, coupled with the cation–π interactions with the edge atoms of graphene, further contribute to the effectiveness of these materials as anode materials. Moreover, $Na^+@F_3H_3$. Gr showed the fastest recovery time, improving the battery capacity through quicker Na-ion insertion and diffusion [59]. Carbon dots (CDs) have also been found to help manage volume fluctuations during lithium alloying and dealloying, contributing to significantly better reversible capacity and longer cycling stability [60]. Some electrode materials undergo substantial volume changes during lithium insertion and extraction, which induce stress and strain, leading to material degradation and electrical disconnection [61, 62]. This results in a reduced coulombic efficiency and a sharp decline in capacity. Owing to their flexibility, these materials help maintain structural integrity, prevent degradation, and enhance performance, reversible capacity, and cycle life. Zhu et al. synthesized CuO/Cu/graphene quantum dot (GQD) triaxial nanowires (CCGs), which are anode materials for LIBs. The unique soft protection offered by GQDs greatly improved the surface conductivity and stability of the nanowire array structure. GQDs help minimize electrode polarization, enhance charge collection efficiency, and manage volume expansion. Consequently, the triaxial nanowire CCG electrode demonstrates a higher reversible capacity (approximately 760 mAh g^{-1} at 1/3 C) and superior rate performance (approximately 60% capacity retention at 10 C) compared with those of uncoated nanowires [63]. QDs present a highly adaptable platform for advancing the performance of diverse energy storage systems. Their unique electronic properties and tunable sizes have a profound impact on both energy storage efficiency and stability. Figure 3 shows a schematic diagram of various applications of quantum dots in energy storage systems.

3.2 *QDs for Optoelectronics Applications*

Over the last few decades, recent advancements in QD technology have significantly influenced the field of optoelectronics. A landmark achievement was the creation of the first quantum dot light-emitting diodes (QD-LEDs) by Colvin and his team in 1994, which paved the way for the development of novel photonic devices [64]. These early QD-LEDs utilizing CdSe QDs exhibited an external quantum efficiency of merely 0.01%, a performance that is more than three orders of magnitude lower than what modern QD-LEDs can achieve.

The remarkable increase in QD-LED efficiency over the years has been largely due to several technological advancements, including the refinement of quantum dot synthesis methods, the optimization of device designs, and improvements in the characteristics of charge transport layers. These advancements have improved the efficiency of QD-LEDs and have also led to the creation of other important QD-based optoelectronic devices, including photovoltaic cells and photodetectors [65, 66]. Currently, QD-based solar cells are gaining attention as promising technologies in third-generation photovoltaics, providing an efficient and cost-effective

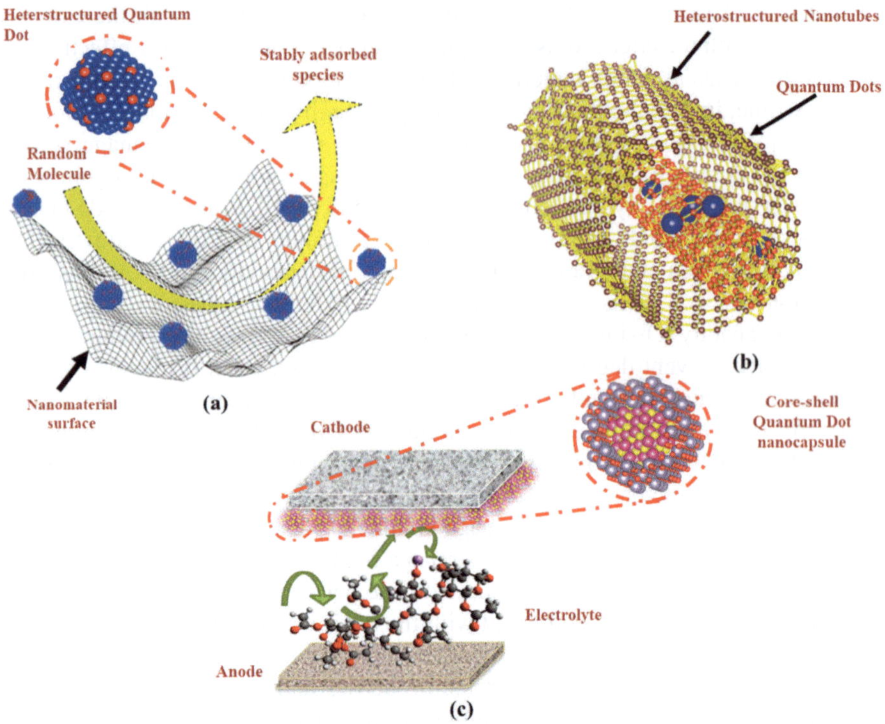

Fig. 3 Schematic diagrams of various applications of QDs. (**a**) QDs as redox mediators converting molecules into stable adsorbed species through QD composite material surfaces, (**b**) QDs as high adsorbers and confinement media for fast electron and ion transfer through heterostructure nanotubes, and (**c**) QDs as promoters of cathode conductivity, electrolyte infiltration and diffusion

approach for energy conversion. Moreover, QD-based photodetectors have demonstrated high effectiveness in near-infrared (NIR) detection, addressing the critical need for sensitivity and precision in this spectral range. The unique semiconductor properties of quantum dots are central to these applications, enabling efficient charge transfer and the formation of various electronic interfaces with metals and insulating materials. Moreover, colloidal quantum dots enable a variety of affordable, solution-based fabrication techniques, making them adaptable to different manufacturing processes [67, 68].

Recent progress in chalcogenide and perovskite colloidal quantum dot (CQD) optoelectronics has led to substantial advancements [69–71]. Over the past decade, CQDs have been extensively researched for numerous applications, including photodetectors, LEDs, lasers, and transistors. The tunable bandgap and solution processability of these materials make them highly attractive for scalable applications. Kirmani et al. designed an enhanced CQD-based optoelectronic device through the development of a mechanistic model for a solid-state ligand exchange technique [72]. By adjusting the concentration, they achieved a 50% decrease in the film thickness, a reduced interdot spacing, an approximately 15 nm redshift in the excitonic absorption peak, and a decrease in the carrier lifetime. The application of optical sensing in the mid- and longwave infrared (MWIR and LWIR) regions has also shown significant potential in areas such as environmental monitoring, gas sensing, hazard detection, and food and product manufacturing inspection. Ramiro et al. investigated intraband absorption and photodetection in heavily doped PbS colloidal QDs within the 5–9 μm wavelength range, surpassing the bandgap of bulk PbS [73]. By utilizing quantum transport equations, they explored the performance of intraband photodetectors, which resulted in enhanced device performance. Piyali et al. investigated monodispersed and environmentally stable cobalt (Co)-doped cadmium sulfide (CdS) QDs [74]. The optical properties revealed the formation of a new doping band near the infrared region, indicating the potential of these materials in optoelectronic applications. The Codoped CdS QDs exhibited enhanced photoconductivity in thin-film heterojunctions, with the photocurrent increasing linearly with increasing Co doping concentration. This enhancement is due to increased electronic transitions resulting from the presence and widening of the doping band, as well as due to a reduction in the bandgap. Specifically, the photocurrent obtained from thin-film heterojunctions with 8% Codoped CdS QDs reached 0.811 mA at an applied voltage of −10 V. Quantum dots also play a crucial role in improving the stability and increasing the photoluminescence quantum yields (PLQYs) of fully inorganic halide perovskite nanocrystals. In a study by Shi et al., cadmium sulfide (CdS) was selected as a shell material to protect $CsPbBr_3$ by synthesizing $CsPbBr_3$@CdS core–shell quantum dots [75]. The results showed that compared with the CsPbBr3@Cs4PbBr6 nanocrystals, the $CsPbBr_3$@CdS QDs exhibit superior stability under various environmental stresses, including thermal, chemical, and light excitation conditions. Furthermore, the PLQY of uncoated $CsPbBr_3$ improved from 81% to 86% when it was encapsulated with a CdS shell. These findings suggest that the CdS shell not only offers enhanced protection but also increases the photoluminescence efficiency of the perovskite core.

3.3 *QDs for Biomedical Applications*

The versatility of QDs has also opened exciting possibilities for their use in biomedical applications such as drug delivery systems, multifunctional theranostic agents, and real-time in vivo imaging, especially in deep tissues. Recent advances in the molecular engineering of QDs have significantly enhanced their functionality in biomedical settings. Researchers have focused on improving the stability of QDs in biological environments, which is crucial for maintaining their functionality over extended periods. Key developments, such as the modification of QD surfaces with biocompatible coatings, have increased their stability, reduced toxicity, and improved their circulation time in the body. These coatings include polymers, peptides, or small molecules that prevent aggregation, minimize immune recognition, and reduce nonspecific binding to biological molecules [76–78]. One of the most promising applications of QDs is in drug delivery systems. Quantum dots can be designed to deliver therapeutic agents directly to specific cells or tissues, enabling targeted therapy while reducing side effects. For example, QDs can be functionalized with targeting molecules that bind to specific cellular receptors that are overexpressed in disease states, such as cancer [79–81]. This targeted approach not only enhances the delivery efficiency of the therapeutic agent but also allows real-time tracking of the drug distribution and release via the fluorescent properties of the QDs. In addition to drug delivery, QDs have shown great potential as multifunctional theranostic agents that combine therapeutic and diagnostic capabilities in a single platform [82]. For example, QDs can be used to simultaneously deliver a drug and monitor its therapeutic effect in real time. This dual functionality is particularly advantageous in cancer treatment, where QDs can be used to deliver chemotherapeutic agents directly to tumors while monitoring the response to treatment through changes in fluorescence signals. This real-time feedback can provide valuable information about the effectiveness of the therapy and allow timely adjustments to the treatment regimen. Another exciting application of QDs is in single-molecule imaging, where they serve as fluorescent probes for tracking the dynamics of individual molecules within cells. QDs offer several advantages over traditional organic dyes for single-molecule imaging, including higher brightness, greater photostability, and the ability to be excited by a wide range of wavelengths. These properties enable long-term imaging of molecular processes at high resolution, providing insights into the fundamental mechanisms of cellular function and disease [83].

4 Conclusion

In summary, nanoscale science explores low-dimensional structures, where materials are confined to dimensions ranging from 1 to 100 nm. These structures consist of quantum wells, quantum wires, and quantum dots, each offering different levels of confinement. As materials are confined, the energy levels of carriers shift from continuous bands in bulk materials to discrete, quantized levels in quantum dots. The Schrödinger equation, which governs the quantum behavior of low-dimensional structures, changes according to the level of confinement. In quantum wires, factors such as the wire dimensions, shape, and characteristics of the surrounding potential barriers affect the energy levels. In quantum dots, where carriers are confined in all three dimensions, the size and shape of the dot determine the energy levels, with confinement energy decreasing as the dot size increases. Excitons, which are bound states of electrons and holes, play a critical role in defining the optical properties of quantum dots. The confinement effects in quantum dots can vary from weak to strong, depending on the dot's size relative to the exciton Bohr radius, which in turn affects the formation of energy levels. A deep understanding of these phenomena is crucial for optimizing the optoelectronic properties of low-dimensional structures, thereby enhancing their applications across various technologies. Quantum dots

(QDs) have revolutionized nanoscience by offering exceptional optoelectronic properties and tunability, which have led to significant advancements in diverse fields, such as energy storage, optoelectronics, and biomedicine. The unique ability of QDs to exhibit size-dependent properties, particularly quantum confinement, has enabled their integration into various high-performance devices and systems. These versatile materials enhance the efficiency and functionality of lithium-ion batteries, supercapacitors, light-emitting diodes, solar cells, and photodetectors, among others.

Acknowledgement The authors would like to acknowledge Tezpur University and Assam Kaziranga University, Assam, India.

References

1. Li, G., Liu, Z., Gao, W., & Tang, B. (2023). Recent advancement in graphene quantum dots based fluorescent sensor: Design, construction and bio-medical applications. *Coordination Chemistry Reviews, 478*, 214966.
2. Zhao, C., Wang, X., Yu, L., Wu, L., Hao, X., Liu, Q., & Lin, X. (2022). Quaternized carbon quantum dots with broad-spectrum antibacterial activity for the treatment of wounds infected with mixed bacteria. *Acta Biomaterialia, 138*, 528–544.
3. Galstyan, V. (2021). "Quantum dots: Perspectives in next-generation chemical gas sensors"–A review. *Analytica Chimica Acta, 1152*, 238192.
4. Facure, M. H. M., Schneider, R., Mercante, L. A., & Correa, D. S. (2020). A review on graphene quantum dots and their nanocomposites: from laboratory synthesis towards agricultural and environmental applications. *Environmental Science: Nano, 7*(12), 3710–3734.
5. Grillot, F., Norman, J. C., Duan, J., Zhang, Z., Dong, B., Huang, H., Chow, W. W., & Bowers, J. E. (2020). Physics and applications of quantum dot lasers for silicon photonics. *Nano, 9*(6), 1271–1286.
6. Bai, Y., Hao, M., Ding, S., Chen, P., & Wang, L. (2022). Surface chemistry engineering of perovskite quantum dots: strategies, applications, and perspectives. *Advanced Materials, 34*(4), 2105958.
7. Ekimov, A. I. (1981). Quantum size effect in three-dimensional microscopic semiconductor crystals. *JETP Letters, 34*, 345.
8. Rossetti, R., Hull, R., Gibson, J. M., & Brus, L. E. (1985). Excited electronic states and optical spectra of ZnS and CdS crystallites in the≊ 15 to 50 Å size range: Evolution from molecular to bulk semiconducting properties. *The Journal of Chemical Physics, 82*(1), 552–559.
9. Calvet, L. E., Chen, J., Lee, T., & Seabaugh, A. (2022). Mark A. Reed (1955–2021). *Nature Nanotechnology, 17*(4), 336–336.
10. Kambhampati, P. (2021). Nanoparticles, nanocrystals, and quantum dots: What are the implications of size in colloidal nanoscale materials? *The Journal of Physical Chemistry Letters, 12*(20), 4769–4779.
11. Melnychuk, C., & Guyot-Sionnest, P. (2021). Multicarrier dynamics in quantum dots. *Chemical Reviews, 121*(4), 2325–2372.
12. Smyder, J. A., & Krauss, T. D. (2011). Coming attractions for semiconductor quantum dots. *Materials Today, 14*(9), 382–387.
13. Nguyen, K. A., Day, P. N., & Pachter, R. (2010). Understanding structural and optical properties of nanoscale CdSe magic-size quantum dots: insight from computational prediction. *The Journal of Physical Chemistry C, 114*(39), 16197–16209.

14. Stręk, W., Makuszewski, K., Jasiorski, M., Lukowiak, E., Bryja, L., Ciorga, M., Sitarek, P., & Misiewicz, J. (1999). Spectroscopic properties of CdS nanoparticles embedded in sol-gel silica glasses. *Optica Applicata, 29*(3), 401–405.
15. Lifshitz, E., Dag, I., Litvin, I., Hodes, G., Gorer, S., Reisfeld, R., Zelner, M., & Minti, H. (1998). Optical properties of CdSe nanoparticle films prepared by chemical deposition and sol–gel methods. *Chemical Physics Letters, 288*(2–4), 188–196.
16. Ekimov, A. I., Efros, A. L., & Onushchenko, A. A. (1985). Quantum size effect in semiconductor microcrystals. *Solid State Communications, 56*(11), 921–924.
17. Rossetti, R., Ellison, J. L., Gibson, J. M., & Brus, L. E. (1984). Size effects in the excited electronic states of small colloidal CdS crystallites. *The Journal of Chemical Physics, 80*(9), 4464–4469.
18. Pütz, N., Veuhoff, E., Heinecke, H., Heyen, M., Lüth, H., & Balk, P. (1985). GaAs growth in metal–organic MBE. *Journal of Vacuum Science & Technology B: Microelectronics Processing and Phenomena, 3*(2), 671–673.
19. Wuister, S. F., Swart, I., van Driel, F., Hickey, S. G., de Mello, C., & Donega. (2003). Highly luminescent water-soluble CdTe quantum dots. *Nano Letters, 3*(4), 503–507.
20. Dagtepe, P., Chikan, V., Jasinski, J., & Leppert, V. J. (2007). Quantized growth of CdTe quantum dots; observation of magic-sized CdTe quantum dots. *The Journal of Physical Chemistry C, 111*(41), 14977–14983.
21. Yamaguchi, K. Y. K., Yujobo, K. Y. K., & Kaizu, T. K. T. (2000). Stranski-Krastanov growth of InAs quantum dots with narrow size distribution. *Japanese Journal of Applied Physics, 39*(12A), L1245.
22. Micic, O. I., Curtis, C. J., Jones, K. M., Sprague, J. R., & Nozik, A. J. (1994). Synthesis and characterization of InP quantum dots. *The Journal of Physical Chemistry, 98*(19), 4966–4969.
23. Pons, T., Pic, E., Lequeux, N., Cassette, E., Bezdetnaya, L., Guillemin, F., Marchal, F., & Dubertret, B. (2010). Cadmium-free CuInS2/ZnS quantum dots for sentinel lymph node imaging with reduced toxicity. *ACS Nano, 4*(5), 2531–2538.
24. Chen, C.-W., Dai-Ying, W., Chan, Y.-C., Lin, C. C., Chung, P.-H., Hsiao, M., & Liu, R.-S. (2015). Evaluations of the chemical stability and cytotoxicity of CuInS2 and CuInS2/ZnS core/shell quantum dots. *The Journal of Physical Chemistry C, 119*(5), 2852–2860.
25. Yang, Y., Lan, J., Zhigang, X., Chen, T., Zhao, T., Cheng, T., Shen, J., Lv, S., & Zhang, H. (2014). Toxicity and biodistribution of aqueous synthesized ZnS and ZnO quantum dots in mice. *Nanotoxicology, 8*(1), 107–116.
26. Xiang, W., Xie, C., Wang, J., Zhong, J., Liang, X., Yang, H., Luo, L., & Chen, Z. (2014). Studies on highly luminescent AgInS2 and Ag–Zn–In–S quantum dots. *Journal of Alloys and Compounds, 588*, 114–121.
27. Long, Z., Zhang, W., Tian, J., Chen, G., Liu, Y., & Liu, R. (2021). Recent research on the luminous mechanism, synthetic strategies, and applications of CuInS2 quantum dots. *Inorganic Chemistry Frontiers, 8*(4), 880–897.
28. Kamel, O. A., Fouad, M., & Ali, M. (2023). A review, water-soluble CuInS quantum dots, strategies and photoluminescence. *International Journal of Nanoscience, 22*(01), 2230005.
29. Vorontsov, D., Fucikova, A., Dedic, V., & Valenta, J. (2023). Cd-free photoluminescent composites based on the ternary chalcogenides quantum dots. *Optical Materials, 143*, 114208.
30. Chi, W., & Banerjee, S. K. (2022). Application of perovskite quantum dots as an absorber in perovskite solar cells. *AngewandteChemie International Edition, 61*(9), e202112412.
31. Zhang, Z., Dahal, N., Ke, X., Choi, D., Yang, E.-H., & Park, J.-R. (2010). Electrochemical characterization of tin quantum dots grown on a carbon nanotube mat as an anode of batteries for medical applications. *Nanoscience and Nanotechnology Letters, 2*(2), 86–88.
32. Gutowski, P., Sankowska, I., Karbownik, P., Pierścińska, D., Serebrennikova, O., Morawiec, M., Pruszyńska-Karbownik, E., Gołaszewska-Malec, K., Pierściński, K., Muszalski, J., & Bugajski, M. (2017). MBE growth of strain-compensated InGaAs/InAlAs/InP quantum cascade lasers. *Journal of Crystal Growth, 466*, 22–29.

33. Heinecke, H., Baur, B., Höger, R., & Miklis, A. (1990). Growth of high purity InP by metalorganic MBE (CBE). *Journal of Crystal Growth, 105*(1–4), 143–148.
34. Hasan, S. M. N., You, W., Ghosh, A., Sadaf, S. M., & Arafin, S. (2023). Selective area epitaxy of GaN nanostructures: MBE growth and morphological analysis. *Crystal Growth & Design, 23*(6), 4098–4104.
35. Kumar, A., Kumar, D., Saikia, M., & Subramania, A. (2023). A review on plant derived carbon quantum dots for bio-imaging. *Materials Advances, 4*(18), 3951–3966.
36. Sobhanan, J., Rival, J. V., Anas, A., Shibu, E. S., Takano, Y., & Biju, V. (2023). Luminescent quantum dots: Synthesis, optical properties, bioimaging and toxicity. *Advanced Drug Delivery Reviews, 197*, 114830.
37. Hashim, H. S., Fen, Y. W., Omar, N. A. S., Daniyal, W. M. E. M. M., Fauzi, N. I. M., Abdullah, J., & Mahdi, M. A. (2024). Surface plasmon resonance sensor based on gold-graphene quantum dots thin film as a sensing nanomatrix for phenol detection. *Optics & Laser Technology, 168*, 109816.
38. Vatankhah, C., Saki, M., & Jafargholinejad, S. (2015). Theoretical and experimental investigation of quantum confinement effect on the blue shift in semiconductor quantum dots. *Oriental Journal of Chemistry, 31*(2), 907–912.
39. Mohammadi, R., Naderi-Manesh, H., Farzin, L., Vaezi, Z., Ayarri, N., Samandari, L., & Shamsipur, M. (2022). Fluorescence sensing and imaging with carbon-based quantum dots for early diagnosis of cancer: A review. *Journal of Pharmaceutical and Biomedical Analysis, 212*, 114628.
40. Chen, L.-L., Zhao, L., Wang, Z.-G., Liu, S.-L., & Pang, D.-W. (2022). Near-infrared-II quantum dots for in vivo imaging and cancer therapy. *Small, 18*(8), 2104567.
41. Huang, X., & Tang, M. (2021). Research advance on cell imaging and cytotoxicity of different types of quantum Dots. *Journal of Applied Toxicology, 41*(3), 342–361.
42. Liu, L., Najar, A., Wang, K., Du, M., & Liu, S. (2022). Perovskite quantum dots in solar cells. *Advanced Science, 9*(7), 2104577.
43. Sahu, A., Garg, A., & Dixit, A. (2020). A review on quantum dot sensitized solar cells: Past, present and future towards carrier multiplication with a possibility for higher efficiency. *Solar Energy, 203*, 210–239.
44. Rahman, M. M., Karim, M. R., Alharbi, H. F., Aldokhayel, B., Uzzaman, T., & Zahir, H. (2021). Cadmium selenide quantum dots for solar cell applications: a review. *Chemistry–An Asian Journal, 16*(8), 902–921.
45. Selopal, G. S., Zhao, H., Wang, Z. M., & Rosei, F. (2020). Core/shell quantum dots solar cells.*Advanced Functional Materials, 30*(13), 1908762.
46. Shaker, M., Ng, S., Ghazvini, A. A. S., Javanmardi, S., Gaho, M. A., Jin, Z., & Ge, Q. (2024). Carbon/graphene quantum dots as electrolyte additives for batteries and supercapacitors: A review. *Journal of Energy Storage, 85*, 111040.
47. Deng, S., Tie, Z., Yue, F., Cao, H., Yao, M., & Niu, Z. (2022). Rational design of ZnMn2O4 quantum dots in a carbon framework for durable aqueous zinc-ion batteries. *Angewandte Chemie, 134*(12), e202115877.
48. Kitai, A. (Ed.). (2008). Luminescent materials and applications. John Wiley & Sons.
49. Xu, Q., Niu, Y., Li, J., Yang, Z., Gao, J., Ding, L., Ni, H., et al. (2022). Recent progress of quantum dots for energy storage applications. *Carbon Neutrality, 1*(1), 13.
50. Liu, Q., Sun, J., Gao, K., Chen, N., Sun, X., Ti, D., Bai, C., Cui, R., & Qu, L. (2020). Graphene quantum dots for energy storage and conversion: from fabrication to applications. *Materials Chemistry Frontiers, 4*(2), 421–436.
51. Wu, P., Xu, Y., Zhan, J., Li, Y., Xue, H., & Pang, H. (2018). The research development of quantum dots in electrochemical energy storage. *Small, 14*(42), 1801479.
52. Roy, E., Nagar, A., Sharma, A., Roy, S., & Pal, S. (2021). Graphene quantum dots and its modified application for energy storage and conversion. *Journal of Energy Storage, 39*, 102606.

53. Kumar, S., Goswami, M., Netrapal Singh, N., Sathish, M. V. R., & Kumar, S. (2022). Exploring carbon quantum dots as an aqueous electrolyte for energy storage devices. *Journal of Energy Storage, 55*, 105522.
54. Veeramani, V., Chen, Y.-H., Wang, H.-C., Hung, T.-F., Chang, W.-S., Wei, D.-H., Shu-Fen, H., & Liu, R.-S. (2018). CdSe/ZnS QD@ CNT nanocomposite photocathode for improvement on charge overpotential in photoelectrochemical Li-O2 batteries. *Chemical Engineering Journal, 349*, 235–240.
55. Kundu, S., Ragupathy, P., & Pillai, V. K. (2016). Effect of reversible lithium ion intercalation on the size-dependent optical properties of graphene quantum dots. *Journal of the Electrochemical Society, 163*(6), A1112.
56. Luo, P., Guan, X., Yu, Y., Li, X., & Yan, F. (2019). Hydrothermal synthesis of graphene quantum dots supported on three-dimensional graphene for supercapacitors. *Nanomaterials, 9*(2), 201.
57. Lijuan, K., Yongqiang, Y., Ruiyi, L., & Zaijun, L. (2016). Phenylalanine-functionalized graphene quantum dot-silicon nanoparticle composite as an anode material for lithium ion batteries with largely enhanced electrochemical performance. *Electrochimica Acta, 198*, 144–155.
58. Li, S., Luo, Z., Hanyu, T., Zhang, H., Deng, W., Zou, G., Hou, H., & Ji, X. (2021). N, S-codoped carbon dots as deposition regulating electrolyte additive for stable lithium metal anode. *Energy Storage Materials, 42*, 679–686.
59. Daryabari, S., Mansouri, S., Beheshtian, J., & Karimkhani, M. (2021). A computational study on the novel defects of graphene quantum dot as a promising anode material for sodium ion battery. *Materials Chemistry and Physics, 265*, 124484.
60. Song, T.-B., Huang, Z.-H., Niu, X.-Q., Liu, J., Wei, J.-S., Chen, X.-B., & Xiong, H.-M. (2020). Applications of carbon dots in next-generation lithium-ion batteries. *ChemNanoMat, 6*(10), 1421–1436.
61. Wang, K.-X., Li, X.-H., & Chen, J.-S. (2015). Surface and interface engineering of electrode materials for lithium-ion batteries. *Advanced Materials, 27*(3), 527–545.
62. Zhou, L., Zhang, K., Hu, Z., Tao, Z., Mai, L., Kang, Y.-M., Chou, S.-L., & Chen, J. (2018). Recent developments on and prospects for electrode materials with hierarchical structures for lithium-ion batteries. *Advanced Energy Materials, 8*(6), 1701415.
63. Chao, D., Zhu, C., Xia, X., Liu, J., Zhang, X., Wang, J., ... & Fan, H. J. (2015). Graphene quantum dots coated VO2 arrays for highly durable electrodes for Li and Na ion batteries. *Nano letters, 15*(1), 565–573.
64. Colvin, V. L., Schlamp, M. C., & Alivisatos, A. P. (1994). Light-emitting diodes made from cadmium selenide nanocrystals and a semiconducting polymer. *Nature, 370*(6488), 354–357.
65. Han, S., Chen, G., Shou, C., Peng, H., Jin, S., & Tu, C. C. (2020). Visibly transparent solar windows based on colloidal silicon quantum dots and front-facing silicon photovoltaic cells. *ACS Applied Materials & Interfaces, 12*(39), 43771–43777.
66. Ba, K., & Wang, J. (2022). Advances in solution-processed quantum dots based hybrid structures for infrared photodetector. *Materials Today, 58*, 119–134.
67. Zhao, Q., Han, R., Marshall, A. R., Wang, S., Wieliczka, B. M., Ni, J., Zhang, J., et al. (2022). Colloidal quantum dot solar cells: progressive deposition techniques and future prospects on large-area fabrication. *Advanced Materials, 34*(17), 2107888.
68. Ouellette, O., Lesage-Landry, A., Scheffel, B., Hoogland, S., García de Arquer, F. P., & Sargent, E. H. (2020). Spatial collection in colloidal quantum dot solar cells. *Advanced Functional Materials, 30*(1), 1908200.
69. Tian, Y., Luo, H., Chen, M., Li, C., Kershaw, S. V., Zhang, R., & Rogach, A. L. (2023). Mercury chalcogenide colloidal quantum dots for infrared photodetection: from synthesis to device applications. *Nanoscale, 15*(14), 6476–6504.
70. Yuan, M., Wang, X., Chen, X., He, J., Li, K., Song, B., Hu, H., et al. (2022). Phase-Transfer Exchange Lead Chalcogenide Colloidal Quantum Dots: Ink Preparation, Film Assembly, and Solar Cell Construction. *Small, 18*(2), 2102340.

71. Lim, S., Lee, G., Han, S., Kim, J., Yun, S., Lim, J., Pu, Y.-J., et al. (2021). Monodisperse perovskite colloidal quantum dots enable high-efficiency photovoltaics. *ACS Energy Letters, 6*(6), 2229–2237.
72. Kirmani, A. R., Walters, G., Kim, T., Sargent, E. H., & Amassian, A. (2020). Optimizing solid-state ligand exchange for colloidal quantum dot optoelectronics: How much is enough?. *ACS Applied Energy Materials, 3*(6), 5385–5392.
73. Ramiro, I., Özdemir, O., Christodoulou, S., Gupta, S., Dalmases, M., Torre, I., & Konstantatos, G. (2020). Mid-and long-wave infrared optoelectronics via intraband transitions in PbS colloidal quantum dots. *Nano letters, 20*(2), 1003–1008.
74. Maity, P., Kumar, S., Ravi Kumar, S. N., Jha, D., Bhattacharyya, S. R., Barman, S. C., Pal, B. N., & Ghosh, A. K. (2021). Role of cobalt doping in CdS quantum dots for potential application in thin film optoelectronic devices. *The Journal of Physical Chemistry C, 125*(3), 2074–2088.
75. Shi, J., Ge, W., Zhu, J., Saruyama, M., & Teranishi, T. (2020). Core–shell CsPbBr3@ CdS quantum dots with enhanced stability and photoluminescence quantum yields for optoelectronic devices. *ACS Applied Nano Materials, 3*(8), 7563–7571.
76. Filali, S., Pirot, F., & Miossec, P. (2020). Biological applications and toxicity minimization of semiconductor quantum dots. *Trends in biotechnology, 38*(2), 163–177.
77. Duo, X., Xu, Q., Li, C., Meng, X., & Feng, Y. (2024). CAG peptide functionalized graphene quantum dots-cationic polymer composite gene carriers. *Journal of Materials Chemistry B, 12*(30), 7357–7366.
78. Ma, Z., Sun, Y., Xie, J., Li, P., Qiujun, L., Liu, M., Yin, P., Li, H., Zhang, Y., & Yao, S. (2020). Facile preparation of MnO2 quantum dots with enhanced fluorescence via microenvironment engineering with the assistance of some reductive biomolecules. *ACS Applied Materials & Interfaces, 12*(13), 15919–15927.
79. Cotta, M. A. (2020). Quantum dots and their applications: what lies ahead?. *ACS applied nano materials, 3*(6), 4920–4924.
80. Ruzycka-Ayoush, M., Kowalik, P., Kowalczyk, A., Bujak, P., Nowicka, A. M., Wojewodzka, M., Kruszewski, M., & Grudzinski, I. P. (2021). Quantum dots as targeted doxorubicin drug delivery nanosystems in human lung cancer cells. *Cancer Nanotechnology, 12*(1), 8.
81. Iannazzo, D., Celesti, C., & Espro, C. (2021). Recent advances on graphene quantum dots as multifunctional nanoplatforms for cancer treatment. *Biotechnology journal, 16*(2), 1900422.
82. Alavi, M., Webster, T. J., & Li, L. (2022). Theranostic safe quantum dots for anticancer and bioimaging applications. *Micro Nano Bio Aspects, 1*(2), 1–11.
83. Wu, J., Shi, Z., Zhu, L., Li, J., Han, X., Xu, M., Hao, S., Fan, Y., Shao, T., Bai, H. and Peng, B. (2022). The design and bioimaging applications of NIR fluorescent organic dyes with high brightness. *Advanced Optical Materials, 10*(8), 2102514.

Applications of Quantum Dots (QDs) in Optical Imaging

Varad Mahesh Phadke, Nirmal Mazumder, and Ajeetkumar Patil

1 Introduction

Bioimaging via optical techniques plays a significant role in biological research. Optical imaging can be used to study a wide range of biological samples containing cell or tissue samples through in vivo or in vitro imaging. It is also used in a broad range, from submicron sizes to macroscopic species and samples. Various optical principles and microscopic methods are used to enhance the contrast and utilize it for bioimaging.

In the last decade, organic fluorescent dyes and proteins have been proven to be vital in bioimaging applications. However, with the progress in technologies related to nanoparticles, ideal ingredients with innovative fluorescent features are being used in diagnosis and bioimaging.

QDs are fluorescent nanoparticles made up of a core bound inside the shell of another semiconductor material with a diameter of 2–10 nm. The size of QDs affects their optical properties, such as quantum yield, absorbance, and photoluminescence [1]. QDs are increasingly being used for detection and labelling in bioimaging. Additionally, QDs have various advantages over traditional dyes and proteins. They are comparably photostable and brighter and have wide excitation and narrow emission spectra in addition to large Stoke shifts. Owing to their exclusive spectral and optical properties, QDs are excellent probes for multicolor imaging [2]. These

V. M. Phadke · A. Patil (✉)
Manipal Institute of Applied Physics, Manipal Academy of Higher Education, Manipal, India
e-mail: ajeetkumar.p@manipal.edu

N. Mazumder
Department of Biophysics, Manipal School of Life Sciences, Manipal Academy of Higher Education, Manipal, Karnataka, India

S. P. Kabekkodu et al. (eds.), *Functional Quantum Dots*, Lecture Notes in Nanoscale Science and Technology 40,
https://doi.org/10.1007/978-3-031-98219-4_2

features provide applications for systems for multiple analyses, such as multicolor staining of cells and tissues and tracking of single-QDs in cells.

Currently, several researchers have investigated the potential applications of QDs (Fig. 1), especially in biomedical treatment and diagnosis. Because of their high quantum yields and absorption coefficients, QDs usually do not need high-intensity excitation beams, so it is easy to avoid photodamaging biological samples.

The variable spectral properties are dependent on the configuration and core size, which results in a wide range of optical fluorescence. Compared with organic dyes, quantum dots (QDs) provide greater photostability along with wide excitation and narrow emission spectra with large stock shifts. Additionally, QDs have high quantum yields and absorption coefficients, so there is no need for high-intensity excitation beams, which prevents the photodamaging of biological samples [3].

Additionally, QDs with emission spectra in the NIR region are capable of both in vitro and in vivo imaging of cells and tissues [4]. Quantum dots (QDs) are promising fluorophores for imaging because of their high photostability, inherent brightness, narrow emission band, and excitability at various wavelengths [4–6]. QDs are composed of many atoms that act as single semiconductor materials such as nanocrystals. These properties differ according to their size, which is 1–10 nm in diameter, whereas smaller QDs (generally having diameters of 5–6 nm) emit short wavelengths. The electrons in the QDs are restricted to distinct energy levels, which are dependent on the particle size. Thus, a smaller particle size corresponds to a larger band gap. Therefore, larger QDs (generally 2–3 nm) emit longer wavelengths

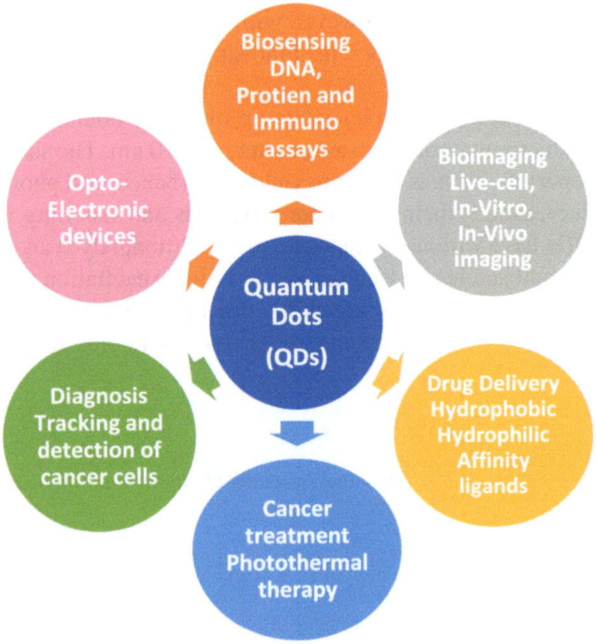

Fig. 1 Schematic of QD applications in various fields

[7]. The crystal-like core of QDs is made up of cadmium selenide and layered with zinc sulfide. Additionally, some QDs are layered with various polymers and molecules, which makes them water soluble and couples various biomolecules to obtain specific functionalities [8, 9]. Considering the optical and structural properties of QDs, fluorescent proteins and organic dyes, QDs overcome most of the disadvantages of dyes and proteins [10]. Table 1 shows the optical and structural features of the QDs compared with those of organic dyes.

The application of QDs in optical imaging has been established for the measurement of oxygen, chloride ion, calcium and enzymatic activity in the cell. QD conjugates, which have temperature- and pH-dependent PL intensities and lifetimes, have been used as intracellular nanothermometers and pH meters [11, 12]. As we consider organic dyes, the neighboring environment, such as polarity, viscosity, ionic strength, and pH, affects the intensity of the absorption and emission bands along with the fluorescence lifetime [3]. Additionally, in comparison with QDs, fluorophores such as organic dyes provide discrete and slight excitation spectra adjacent to their wide-ranging emission spectra, which makes it difficult to resolve two or more dyes [13]. For multimodal advanced fluorescence imaging, fluorophores should possess features such as intense photoluminescence, high quantum yield, biocompatibility in the case of in vivo studies, the ability to modify the surface, photochemical stability, and reproducibility [14]. These restrictions can be overcome by using nanocrystals such as QDs. They have the exclusive properties of being size tunable and having the ability to produce high fluorescence emission and photostability, which show superiority over organic dyes and proteins [15], especially considering multimodal and advanced fluorescence imaging and bioimaging. Additionally, QDs can be associated with several molecules, i.e., proteins, DNA and antibodies, so they have a widespread range of applications in bioscience and bioimaging. QDs are used to confine mRNA and proteins in cells, label cancer markers, and track embryonic stem cells in deep tissues [16].

Generally, biomedical applications require water-soluble and stable QDs that are functionalized by biomolecules such as peptides, proteins, drugs, and antibodies. However, as functionalized QDs have better optical properties, they have several drawbacks, such as their complex surface chemistry and comparatively bulky size

Table 1 Comparison of quantum dots (QDs) with organic dyes

Sr. no.	Property	Organic dyes	Quantum dot (QDs)
1.	Size	≤1 nm	2–10 nm
2.	Emission spectrum	Asymmetric IR, UV region	Symmetric and Gaussian curve IR—UV region, varies by size and pH of QDs
4.	Absorption spectra	Discrete	Gradual increase towards UV wavelengths
5.	Quantum yield	0.5–1.0	0.1–0.8
6.	Fluorescence lifetime	1–10 nm	10–100 nm
7.	Photostability	Less	High
8.	Stock shift	50–150 nm	≤50 nm
9.	Thermal stability	Low	High

[17, 18]. In clinical use, cell toxicity and nonbiodegradability are the main disadvantages of QDs. Therefore, more improvements are needed in the design of biocompatible, nonhazardous probes based on QDs.

2 Fluorescence

Fluorescence is the process in which incident photons are absorbed by fluorophores in a sample and fluorescence light is emitted. The energy of an emitted photon is lower than that of the incident photon. Therefore, incident photons have higher wavelengths than emitted photons. This change in wavelength is called Stokes's shift. On the basis of this phenomenon, a fluorescence microscope can detect fluorescent radiation in image or light form.

2.1 Jablonski Diagram

During absorption, a photon with higher energy is absorbed by an electron in the ground state (S_0), and it travels to the higher energy state. Vibrational relaxation is the energy loss between vibrational energy levels, which takes place in nonradiative form. This spare vibrational energy is exerted as kinetic energy to the other vibrational modes of the same or different molecules. In fluorescence, the spin state of the electron relaxes and returns to the ground state by emitting a photon. Internal conversion is nonradiative emission in which an electron in a higher energy excited state travels to the lower energy excited state. There is no change in the electron spin, and not a single photon is emitted throughout the transition. Intersystem crossing is one more type of nonradiative emission. In contrast to internal conversion, there is a change in the spin state of the excited electron. Phosphorescence is a form of fluorescence where an electron relaxes via the emission of a photon (Fig. 2).

2.2 Fluorophores or Fluorochromes

The fluorophores absorb energy at a specific wavelength and emit light at a longer wavelength. The wavelength absorption, efficiency to transfer energy, and time taken for emission are dependent on the structural and chemical properties of the fluorophore.

The principal characteristics of fluorophores are as follows:

- Maximum excitation and emission wavelengths (nm): These wavelengths correspond to the highest values in the excitation and emission spectra (generally one peak from both spectra).

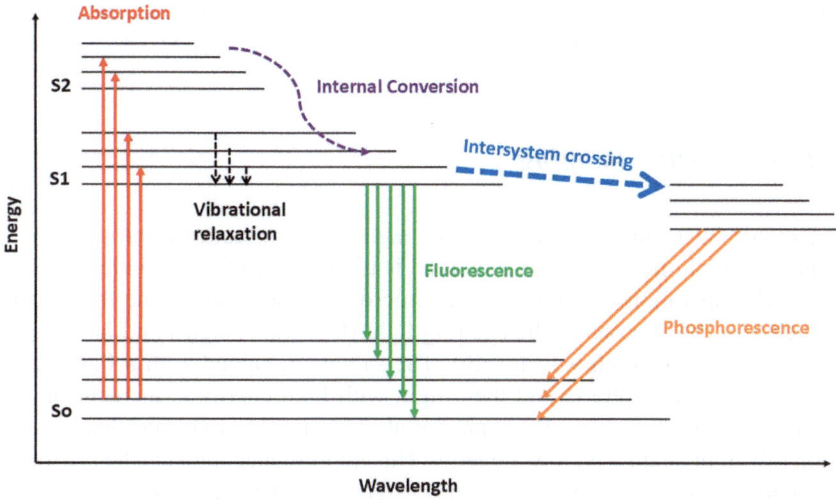

Fig. 2 Schematic of the Jablonski diagram

- Molar absorption coefficient ($\text{Molar}^{-1}\ \text{cm}^{-1}$): At a particular wavelength, we can study the dependence of the quantity of light absorbed and the fluorophore concentration in the solution.
- Quantum yield: Ability of the energy transfer from incident light to emitted light, i.e., the total number of photons required for emission per total number of absorbed photons.
- Lifetime: Time for which the electron is present in the excited state of a fluorophore before travelling back to the ground state.
- Stoke shift: the difference between the maximum excitation and emission wavelengths.
- Dark fraction: the amount of the active molecules throughout the emission.
- Photobleaching or photoresistance refers to the loss of fluorescence intensity due to continuous high-intensity light excitation.

The fluorophores are classified into two main categories: endogenous fluorophores and exogenous fluorophores. Endogenous fluorophores are biomolecules that naturally exist in cells. These can produce autofluorescence. However, autofluorescence caused by endogenous fluorophores creates undesirable noise in the background [19]. There are several biological structures and processes for which images are difficult to obtain via the use of endogenous fluorophores.

Additionally, some cells or tissues do not have autofluorescence; in these cases, labelling of these biological structures with exogenous fluorophores is required for bioimaging. Some fluorophores can be stained (dispersed) throughout the cell, whereas others can be concentrated in a specific portion of the cell. In some cases, a fluorophore is chemically fixed for targeting a specific biological site, a specific type of cancer cell, or a single organelle in the cell. Therefore, the sensitivity and specificity of a fluorophore can be used to obtain biochemical and structural information about cells and tissues [19, 20].

2.3 Fluorescence Imaging

It is an important technique used for bioimaging. Fluorescence microscopy is a technique in which the fluorescence method is used to excite a molecule and emit radiation. The operation of a fluorescence microscope depends on the fluorescent dye, which absorbs photons of the exciting light (low wavelength) and later transmits fluorescent light (high wavelength). The sample is tagged with a fluorescent material called a fluorophore, and then, light with high energy is incident on the sample through the lens. The fluorophores absorb incident light and lead to the emission of a light beam with lower energy. This fluorescence light can be separated from neighboring radiation by using filters designed for that precise wavelength and travels towards the detector.

Initially, the microscope has a filter that only permits radiation with a definite wavelength that matches the fluorescence material. The electron is excited by the incident light and moves to a higher energy level. After some time, those electrons travel back to the lower energy level by emitting a light beam. While detecting it should be visible to human eyes, as emitted light has lower energy and longer wavelengths than incident light does, the emitted light beam passes through a second filter,

2.4 Epifluorescence Illumination

Epifluorescence illumination is a special setup of optical components that allows specific illumination of a sample. The objective lens acts as a source of light that excites the fluorophore, and the emitted fluorescent light is collected through the microscope. Compared with other arrangements of fluorescence microscopy, epifluorescence illumination is advantageous because it blocks a small amount of emitted light. Additionally, the excitation and emission light overlap, and the dichroic mirror supports the discretization of the excited and emitted beams.

The various microscopy techniques used in bioimaging are as follows:

2.5 Widefield Fluorescence Microscopy

Widefield fluorescence microscopy is a basic and general technique in biology and biomedical research. In this technique, the light beam illuminates the complete sample to excite the fluorophore in the sample. Mercury or xenon lamps are generally used as the excitation source. An innovative method for excitation light is the use of intense single-wavelength LEDs (Fig. 3).

As discussed in terms of advantages, the whole region of the sample is illuminated immediately, which helps with rapid imaging. Overall, to avoid photobleaching and phototoxicity, low-intensity imaging is used in wide-field imaging. Generally, the widefield imaging method is used to observe samples in real time without any camera or detector, which is advantageous for the operator in selecting the exact region of the sample. In particular, if a sample is kept under illumination for a longer time after manual observation, it is necessary to minimize photobleaching by reducing the observation time. Additionally, as advantageous parts, wide-field microscopes are less expensive, have simple setups and multipurpose and flexible arrangements. In contrast, there are several disadvantages, including the lower resolution of the images and the existence of uneven contrast because of the irregular alignment and illumination of the optical system.

For widefield microscopy applications, QDs have several advantages over dyes, namely, higher photostability and larger absorption cross-sections. In addition to recording images of cellular structures at high resolution, it is very important to study molecular interactions. Single-molecule tracking (SMT) generally tracks a molecule, whereas single-particle tracking (SPT) studies the dimensional particles bound to the molecular group. The advantage of QDs over dyes and proteins is that QDs have narrow spectra with high SNRs, permitting improved spatial resolutions of approximately 5–10 nm [21]. One of the researcher groups, Walter et al., revised various SPT techniques and recommended a standard and conventional wide-field microscope, as the autofocus function is very advantageous for maintaining a uniform distance of the sample [21].

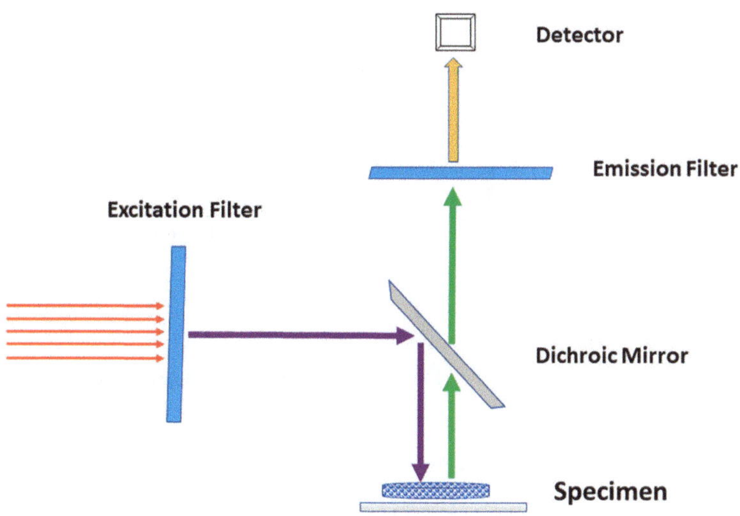

Fig. 3 Schematic of a widefield microscope

2.6 *Confocal Fluorescence Microscopy*

Compared with other conventional microscopes, which produce blurred images, confocal microscopy produces sharp images of objects. Confocal microscopes include only the light beam from the object, which is from the focal plane of the microscope, so they exclude most of the light coming from the object. Additionally, the obtained images have less haze along with better contrast. Therefore, for better observation of fine details, this technique also allows the construction of three-dimensional (3D) images of the object [22, 23].

Figure 4. The setup of confocal microscopy. The coherent light beam is used as an excitation source, which is emitted by the laser source. The incident light travels through a pinhole located in the conjugate plane, pointing towards the sample [24]. Another pinhole is situated very opposite to the detector or a photomultiplier tube (PMT). As a dichromatic mirror reflects the laser and is scanned by the sample, a secondary beam emitted from the sample travels and returns through a dichromatic mirror by focusing towards the detector [25].

The extra amount of fluorescence light in the upper and lower parts of the focal plane is not concentrated on the pinhole, as it is an out-of-focus beam. As only a small amount of out-of-focus fluorescence light passes through the aperture, the maximum amount of extra light is enabled for detection by PMT, which does not contribute to the resulting image.

In conventional widefield microscopy, the whole sample is exposed to an intense light source, and we can obtain the resulting image with secondary fluorescence

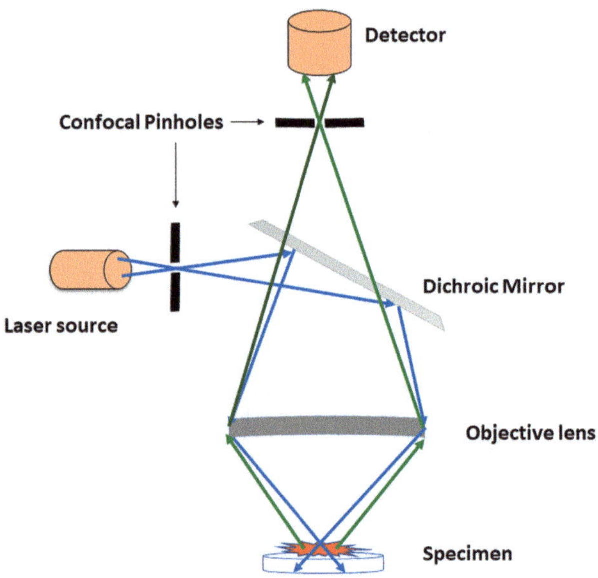

Fig. 4 Schematic of the confocal microscope

emission through eyepieces or detectors. However, there is a different mechanism for image development under a confocal microscope [26]. Additionally, the confocal microscope involves several laser sources used for excitation, along with optical and electronic apparatuses; a detector, i.e., a photomultiplier tube; and a computational setup for processing, acquisition, and analysis of the images.

A multipurpose probe is dependent on bright cadmium-free CISe/ZnS QDs coupled with the CGKRK peptide, which supports tumor targeting. These QDs showed higher photostability, a sensibly long circulation half-life, and minimum toxicity in addition to strong tumor-specific homing. QDs, which are applied to cancer cells, enable noninvasive real-time imaging of the tumor-homing process for a longer time [27]. Additionally, they allow strong intertumoral signals with TP imaging and allow single-cell tracking at the invasion front of glioblastoma. The NIR, TG, and TP imaging results verified the appropriateness of these derived QDs for multimodal imaging, which can increase the consistency of imaging by eliminating artefacts and improving the sensitivity and resolution. As the spectra of QDs are tunable, multiplexed imaging is also possible. These properties suggest the widespread applicability of target-bound QDs in biomedical research and imaging of human environments via optical systems [28].

Another study was conducted by Low-Nam et al., in which CdSe/ZnS QDs were encapsulated with ferric oxide in phospholipid micelles. The cellular uptake of QD micelles by cancer cells was studied via confocal microscopy. During the first 24 h of incubation with micelles liquefied in HPLC water, the QD micelles were concentrated throughout the cell membrane and cytoplasm, and the fluorescence intensity was high. Similarly, after 72 h of continuous cultivation, the QD micelles provided good fluorescence intensity [29]. Figure 5 shows the cellular uptake of the QDs in the HeLa cells.

After 72 h of incubation, for verification of the cellular uptake of the QD micelles, the samples were harvested via a trypsin–EDTA solution and then reseeded. The additional 72 h timeline also resulted in good fluorescence intensity from the cells. Therefore, the results of this study indicate that QD micelles are extremely stable for cancer cells and maintain their fluorescence properties [29].

Additionally, in the detection of cervical cancer cells, QDs can play a vital role. The identification of cancer markers, such as EGFR and QDs, allows the targeting and interrogation of these markers via optical labels [30]. Imaging was carried out via laser scanning confocal microscopy (LSCM) at excitation wavelengths of 488 nm and 568 nm. In the case of anti-EGFR antibodies, images of SiHa cells with precise targeting exhibited stronger intensities at both 488 and 568 nm than nonspecific IgG antibodies did. Primary outcomes show that QDs coupled with anti-EGFR antibodies can be used along with advanced optical technologies to study molecular-level variations in cervical cancer [31].

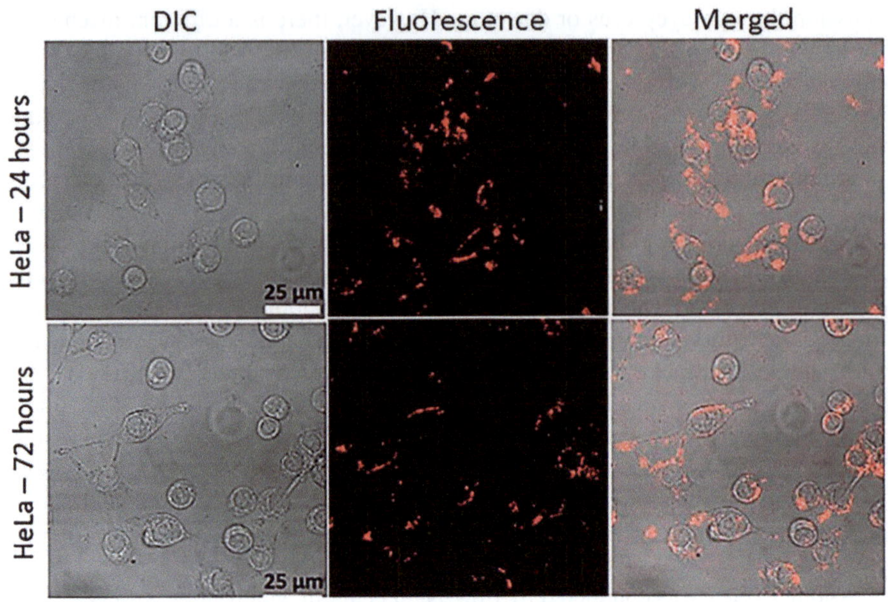

Fig. 5 Cellular uptake of micelles into HeLa cells after incubation for 24 and 72 h [29]

3 Advanced Techniques in Fluorescence Imaging

Owing to their high sensitivity and specificity, fluorescence imaging offers an effective and exceptional approach for studying fixed and living cells. Fluorescence microscopes can detect the fluorescence beam emitted from labelled samples in the form of images, which are used to obtain spectra of intensities and wavelengths. Additionally, such techniques enable the imaging and examination of composite dynamic activities in cells, organelles, and suborganelle components within biological samples.

The various techniques related to fluorescence microscopy are as follows:

- Fluorescence Resonance Energy Transfer (FRET).
- Two-Photon Fluorescence Microscopy (TPM).
- Fluorescence Recovery After Photobleaching (FRAP).
- Fluorescence lifetime imaging microscopy (FLIM).

3.1 Fluorescence Resonance Energy Transfer (FRET)

Energy is transported in a radiation-free form from one chromophore to another chromophore. During this energy transfer, an essential requirement for effective transfer over 10–100 A^0 is that the fluorescence spectrum of D and the absorbance spectrum of An overlap. These values of the quantum yield of D and the absorption coefficient of A are sufficiently high [32].

Additionally, dipole–dipole interactions occur when transition dipoles D and A must be oriented favorably relative to each other or when one (or both) should have a definite degree of rotational freedom. As it is a type of radiation-free transfer, D and A do not emit a photon or absorb a single photon. The energy transfer rate is dependent on the following factors: (1) the degree of spectral overlap between the emission and absorption spectra of D and A, respectively. (2) the quantum yield of D, (3) the positioning of D and A, (4) the transition dipoles, and (5) the total distance between D and A [32]. The efficiency of energy transfer (E) is the number of quanta that are transferred from donor D to acceptor A (Fig. 6). The quantum yield (E) of energy transfer is given as [33]

$$E = \frac{\text{Total amount of quanta transferred from D}}{\text{Total amount of quanta absorbed by D}}$$

The QDs used as donors in FRET are becoming wider with the extensive style of integrating QDs for the analysis of biomolecules and sensors. QD-FRET examination was performed for the first time by Mattoussi and colleagues, in which dye-labelled protein-QDs were used to illuminate QDs in the form of donors and acceptors. QDs coupled with molecules help in the detection of specific targets [6]. In addition, another research team verified that the retrovirus human immunodeficiency virus (HIV) can be labelled over a membrane through the combination of acceptor proteins, such as site-specific biotinylation, and can possibly be detected via QDs coupled with streptavidin, which can be used as markers for the virus. Additionally, they verified that these markers can be helpful for labelling lentiviruses and gamma viruses. Because of their high photostability, QDs detect a single virus [34, 35]. Additionally, Dennis et al. established QD-based sensors that combine photostable QDs and pH-sensitive fluorescent proteins, which increased the photostability and sensitivity to organic fluorescent dyes [36].

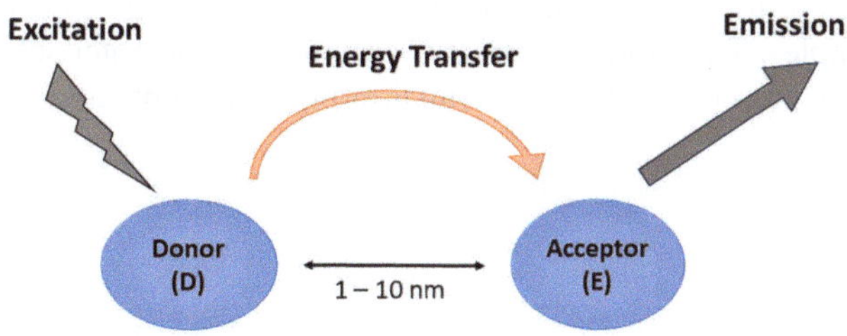

Fig. 6 Energy transfer from the donor to the acceptor in the form of fluorescence

3.2 *Two-Photon Fluorescence Microscopy (TPM)*

In TPM, absorption takes place within a small volume in the focal plane. The absorption plane can be approximately half a micron [37]. Outside this plane, the sample is exposed only to a low average-intensity beam. Thus, photobleaching and phototoxicity are restricted to the plane of focus, and a sample with less potentially damaging radiation makes this technique suitable for live-cell imaging [38]. Additionally, a decrease in photodamage increases the ability to study biological samples. Additionally, in fields such as physiology, neurobiology, embryology, and tissue engineering, where imaging of highly scattering tissue or cells is needed, TPM is an impactful technique [39]. Additionally, two-photon microscopy can be used in noninvasive optical surgery, for which high-speed imaging is required [40, 41]. One of the researchers' groups, Bharali et al., prepared and verified the application in bioimaging by preparing photostable QDs InP-ZnS coupled with folic acid (FA). Bioimaging of QD FA in cancer cells (oral epidermoid carcinoma cells) with a folate receptor (FR) was executed via two-photon and confocal microscopy. Additionally, the QDs were improved by InP to decrease the remaining toxicity. Research groups have mentioned and claimed that the revised QDs provide a high quantum yield during imaging [27]. Wu et al. synthesized nitrogen-doped GQDs coupled with polyethene-imine (PEI) polymers with nitrogen and sulfur atoms. Therefore, the nitrogen–graphene QDs-polymers exhibited high optical luminescence along with high quantum yield. The results suggest the application of prepared GQDs as probes for noninvasive two-photon imaging and for the detection of biological samples via two-photon emission [42].

Zhang et al. examined water-soluble CdS QDs via the TPM imaging technique. For comparison, first, they recorded images of HePG2 cells without incubation with CdS QDs. Figure 7(a1, b1, c1, d1) shows the brightfield image, green and red channel images and a combination of red and green channel images, respectively. The images of label-free cells confirmed the presence of green autofluorescence [43].

Later, the HepG2 cells were incubated with 160 μM CdS QDs for 6 h and washed three times with PBS. Images of the cells were recorded via the TPM imaging technique. Figure 7(a2, b2, c2, d2) shows the brightfield image, green and red channel images and a combination of red and green channel images, respectively. The cells tagged with CdS QDs emitted bright and sharp fluorescence has a red color. This red fluorescence and bright spots confirmed the intracellular uptake of the CdS QDs. The brightness of the emitted fluorescence was 38.2 times greater than that of the label-free samples. This study confirms the application of water-soluble CdS QDs as prominent fluorescent probes for cellular bioimaging applications [44].

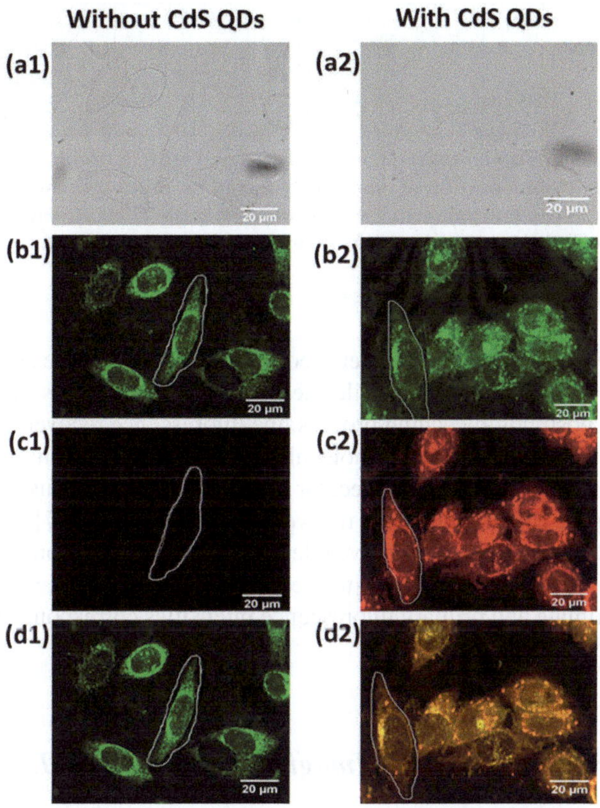

Fig. 7 Brightfield and Two-Photon fluorescence images of HePG2 cells without labelling CdS QDs where (**a1**) Brightfield image. (**b1**) Green channel fluorescence image. (**c1**) Red channel fluorescence image. (**d1**) Combination of red and green channel images. Brightfield and Two-Photon fluorescence images of HePG2 cells with labelling CdS QDs where. (**a2**) Brightfield image. (**b2**) Green channel fluorescence image. (**c2**) Red channel fluorescence image. (**d2**) Combination of red and green channel images [44]

3.3 Fluorescence Recovery After Photobleaching (FRAP)

FRAP is a microscopy-based method used to examine the movement of fluorescent molecules. In FRAP, molecules in an ROI inside the cell are illuminated by high-intensity light. If there are mobile fluorescent species, the lightened molecules in the ROI are replaced by other molecules by diffusion from adjacent regions. The observed intensity retrieval in the ROI provides information about the movable molecules along with the half-time required for recovery [45].

As shown in Fig. 8, FRAP is based on the diffusion rate of the molecule. In a sample labelled with fluorescent probes, the molecules in a particular area are bleached with an intense light source, referred to as photobleaching, as shown in Fig. 8B. This process decreases the intensity of fluorescence in the bleached region,

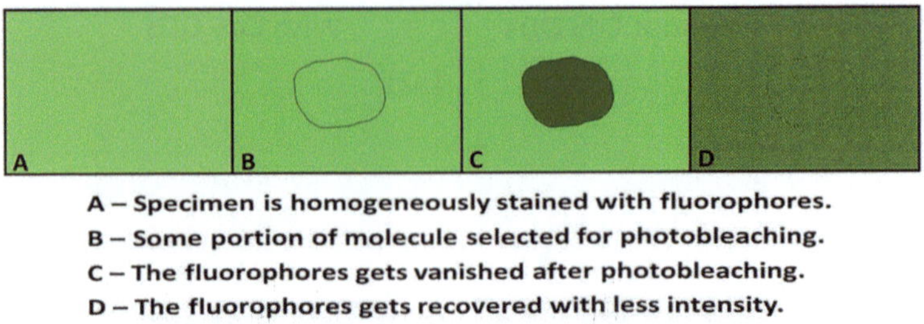

Fig. 8 Representation of the FRAP process

as shown in Fig. 8C. Then, the unbleached fluorophores from the surrounding area diffuse to the bleached section, while the bleached fluorophores diffuse out [46]. Owing to this postphotobleaching process, the fluorescence recovers at a rate that is dependent on the mobility of the molecules. By using a scientific model, we can produce quantitative data about molecular dynamics, such as diffusion and binding, by examining the time required for recovery of fluorescence [47]. If all the molecules in the ROI are completely moveable, then the ultimate fluorescence intensity after bleaching is almost equivalent to the prephotobleaching intensity. The difference in the ultimate and primary fluorescence intensities corresponds to the fraction of the bleached molecules [48].

3.4 Fluorescence Lifetime Imaging Microscopy (FLIM)

This technique provides an image on the basis of the variance in the excited state decay rate of the fluorescent sample. Thus, it is an imaging technique in which contrast is dependent on the lifetime of the fluorophores and does not depend on their emission wavelength. In other words, the fluorescence lifetime is the average time for which a molecule stays in an excited state by the emission of a single photon [32]. The fluorescence lifetime is independent of the concentration, sample thickness, photobleaching, and/or excitation intensity. Moreover, the fluorescence lifetime is also dependent on several chemical parameters, such as pH, the concentration of oxygen ions, and molecular binding [49].

Time-correlated single photon counting (TCSPC) is used to calculate the fluorescence lifetime. In TCSPC, we can measure the time required for the emitted photon to travel up to the detector [50]. TCSPC involves a “start” signal provided by the photodiode and a “stop” signal at the detectors. This delay in time frequently takes place when identifying the data of the emission spectra (Fig. 9) [37].

Huang et al. observed the fluorescence lifetime of CDs through FLIM images (Fig. 10) of HeLa cells. Additionally, phasor FLIM analysis was used to detect the CDs in the lysosomes, cytoplasm, and nucleus of the HeLa cells. In this study, the CDs in the lysosomes had a short lifetime of 1.6 ns. However, in the case of the nucleus, the fluorescence lifetime was significantly longer than that in the

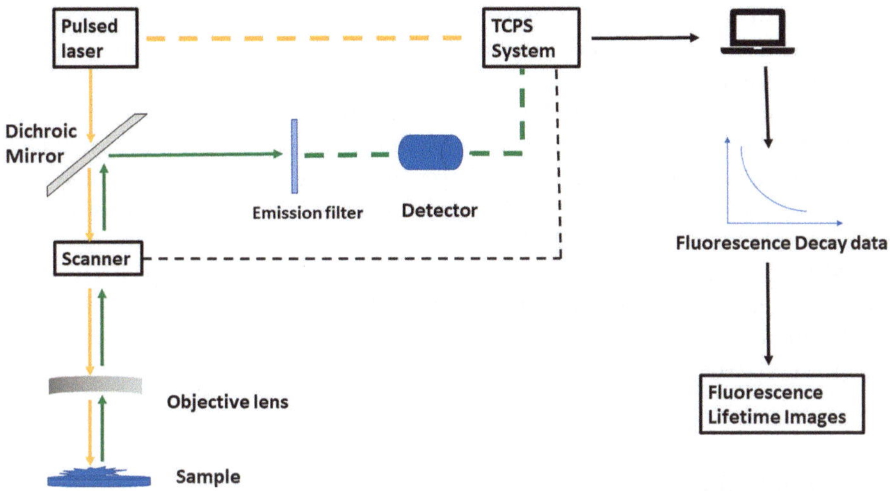

Fig. 9 Schematic of FLIM with TCSPC

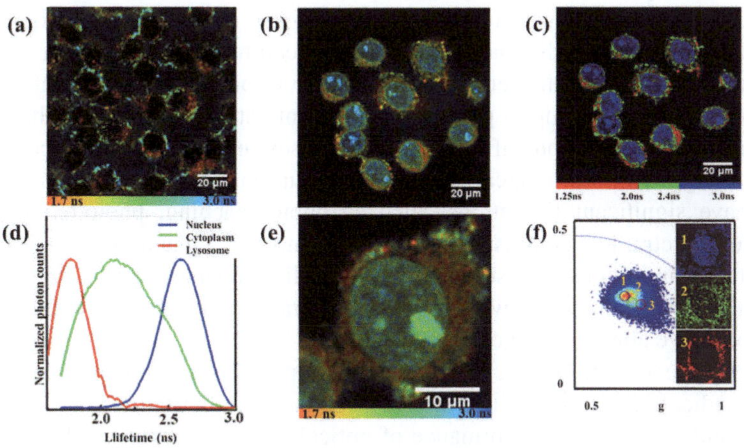

Fig. 10 FLIM images of the HeLa cells (**a**) after incubation with CDs for 2 h. (**b**) After incubation for 2 h and culture in digitonin for 3 min for cell permeabilization. (**c**) Fluorescence lifetimes of the CDs in the lysosomes, cytoplasm, and nucleus. (**d**) Profile of the fluorescence lifetime distribution in the lysosomes, cytoplasm, and nucleus. (**e**) FLIM images of a permeabilized HeLa cell. (**f**) The phasor FLIM distribution for (1) the nucleus, (2) the cytoplasm, and (3) lysosomes [51]

cytoplasm. This study provides a helpful way to use pH-sensitive CDs along with the FLIM technique to examine the intracellular uptake of CDs through composite biological samples [51].

Goreham et al. reported the synthesis of water-soluble, comparably less cytotoxic GOQDs in addition to folic acid as a functional group. These prepared QDs proved to be very less toxic to macrophages, which are among the types of WBCs. Moreover, the same QDs were used to record images of the HaCaT cells via FLIM and multiphoton microscopy (MPM). The researchers stated that, for the first time, MPM-FLIM can be used for the separation of fluorescence formed by GOQDs. The combination of MPM-FLIM delivers a label-free and noninvasive technique to study fluorescent nanoparticles with overlapping spectra, which are difficult to image via conventional methods [52].

4 Conclusion

Optical imaging has become a revolutionary technique in the fields of life sciences and biomedical research, providing molecular and morphological information about biological samples. Optical imaging techniques such as widefield microscopy, confocal microscopy, and two-photon microscopy have enabled researchers to obtain precise images with additional morphological features. As we consider biological samples, most samples are semitransparent and viscous in nature. Therefore, label-free images of these samples provide limited contrast and specificity, which makes further feature extraction difficult. In such cases, exogenous fluorescent probes must be tagged into biological samples. Conventional fluorescent probes such as dyes have significant limitations, such as photobleaching, phototoxicity, broad emission spectra, low contrast, and low quantum yield. Quantum dots (QDs) overcome these limitations with superior photostability, tunable particle sizes, and narrow emission spectra along with enhanced contrast and brightness, making them a transformative alternative in optical imaging.

The application of QDs in optical imaging represents a significant improvement in biomedical research. Their ability to overcome the limitations of conventional probes and enhance the performance of optical imaging techniques has set a new benchmark for sensitivity, resolution, and versatility. This chapter highlights the evolutionary application of QDs in optical imaging for detailed, accurate, and innovative advancements in life science and biomedical research.

Acknowledgement The author of this book chapter, Mr. Varad Mahesh Phadke, is thankful to the Manipal Academy of Higher Education, Manipal, for providing Dr. T.M.A. Pai Scholarship for the research work.

References

1. Qi, L., & Gao, X. (2008). Emerging application of quantum dots for drug delivery and therapy. *Expert Opinion on Drug Delivery, 5*(3), 263–267. https://doi.org/10.1517/17425247.5.3.263
2. Rousserie, G., et al. (2010). Semiconductor quantum dots for multiplexed bio-detection on solid-state microarrays. *Oncology and Hematology, 74*, 1–15. https://doi.org/10.1016/j.critrevonc.2009.04.006
3. Resch-Genger, U., Grabolle, M., Cavaliere-Jaricot, S., Nitschke, R., & Nann, T. (2008). Quantum dots versus organic dyes as fluorescent labels. *Nature Methods, 5*(9), 763–775. https://doi.org/10.1038/nmeth.1248
4. Huang, H., Bai, Y. L., Yang, K., Tang, H., & Wang, Y. W. (2013). Optical imaging of head and neck squamous cell carcinoma in vivo using arginine-glycine-aspartic acid peptide conjugated near-infrared quantum dots. *OncoTargets & Therapy, 6*, 1779–1787. https://doi.org/10.2147/OTT.S53901
5. Gu, L., et al. (2013). In vivo time-gated fluorescence imaging with biodegradable luminescent porous silicon nanoparticles. *Nature Communications, 4*, 77–73. https://doi.org/10.1038/ncomms3326
6. Yang, K., et al. (2010). Quantum dot-based visual in vivo imaging for oral squamous cell carcinoma in mice. *Oral Oncology, 46*(12), 864–868. https://doi.org/10.1016/j.oraloncology.2010.09.009
7. Brus, L. E. (1984). Electron-electron and electron-hole interactions in small semiconductor crystallites: The size dependence of the lowest excited electronic state. *Journal of Chemical Physics, 80*(9), 4403–4409. https://doi.org/10.1063/1.447218
8. Wang, S., Yu, J., Zhao, P., Guo, S., & Han, S. (2021). One-step synthesis of water-soluble CdS quantum dots for silver-ion detection. *ACS Omega, 6*(10), 7139–7146. https://doi.org/10.1021/acsomega.1c00162
9. Shen, L. (2011). Biocompatible polymer/quantum dots hybrid materials: Current status and future developments. *Journal of Functional Biomaterials, 2*(4), 355–372. https://doi.org/10.3390/jfb2040355
10. Schirhagl, R., Chang, K., Loretz, M., & Degen, C. L. (2014). Nitrogen-vacancy centers in diamond: Nanoscale sensors for physics and biology. *Annual Review for Physical Chemistry, 65*, 83–105. https://doi.org/10.1146/annurev-physchem-040513-103659
11. Serhan, M., et al. (2020). Total iron measurement in human serum with a novel smartphone-based assay. *IEEE Journal of Translational Engineering in Health and Medicine, 8*, 1–9. https://doi.org/10.1109/JTEHM.2020.3005308
12. Albers, A. E., Chan, E. M., McBride, P. M., Ajo-Franklin, C. M., Cohen, B. E., & Helms, B. A. (2012). Dual-emitting quantum dot/quantum rod-based nanothermometers with enhanced response and sensitivity in live cells. *Journal of American Chemical Society, 134*(3), 9565–9568. https://doi.org/10.1021/ja302290e
13. Stenspil, S. G., & Laursen, B. W. (2024). Photophysics of fluorescent nanoparticles based on organic dyes—Challenges and design principles. *Royal Society of Chemistry, 15*, 8625–8638. https://doi.org/10.1039/d4sc01352b
14. Luo, M., Yukawa, H., & Baba, Y. (2022). Micro-/nano-fluidic devices and in vivo fluorescence imaging based on quantum dots for cytologic diagnosis. *Royal Society of Chemistry, 22*, 2223–2236. https://doi.org/10.1039/d2lc00113f
15. Medintz, I. L., Mattoussi, H., Clapp, A. R., Clapp, A., & Medintz, I. (2008). Potential clinical applications of quantum dots. *International Journal of Nanomedicine, 3*(2), 151–167. https://doi.org/10.2147/ijn.s614
16. Garon, E. B., Marcu, L., Luong, Q., Tcherniantchouk, O., Crooks, G. M., & Koeffler, H. P. (2007). Quantum dot labelling and tracking of human leukemic, bone marrow and cord blood cells. *Leukemia Research, 31*(5), 643–651. https://doi.org/10.1016/j.leukres.2006.08.006
17. Qian, Z., et al. (2013). Surface functionalization of graphene quantum dots with small organic molecules from photoluminescence modulation to bioimaging applications: An experimen-

tal and theoretical investigation. *RSC Advance, 3*(34), 4571–4579. https://doi.org/10.1039/c3ra42066c

18. Kilina, S. V., Tamukong, P. K., & Kilin, D. S. (2016). Surface chemistry of semiconducting quantum dots: Theoretical perspectives. *Accounts of Chemical Research, 49*, 2127–2135. https://doi.org/10.1021/acs.accounts.6b00196
19. Wizenty, J., et al. (2020). Recent advances and the potential for clinical use of autofluorescence detection of extra-ophthalmic tissues. *Molecules, 25*, 2095–2128. https://doi.org/10.3390/molecules25092095
20. Hickey, S. M., et al. (2022). Fluorescence microscopy—An outline of hardware, biological handling, and fluorophore considerations. *Cells, 11*, 35–67. https://doi.org/10.3390/cells11010035
21. Pons, T., & Mattoussi, H. (2009). Investigating biological processes at the single molecule level using luminescent quantum dots. *Annual Biomedical Engineering, 37*(10), 1934–1959. https://doi.org/10.1007/s10439-009-9715-0
22. Jonkman, J., & Brown, C. M. (2015). Any way you slice it—A comparison of confocal microscopy techniques. *Journal of Biomolecular Techniques, 26*(2), 54–65. https://doi.org/10.7171/jbt.15-2602-003
23. Singh, V., Sharma, N., & Singh, S. (2020). A review of imaging techniques for plant disease detection. *Artificial Intelligence in Agriculture, 4*, 229–242. https://doi.org/10.1016/j.aiia.2020.10.002
24. Merchant, F. A., & Diaspro, A. (2022). Three-dimensional imaging. In *Microscope image processing* (2nd ed., pp. 247–317). https://doi.org/10.1016/B978-0-12-821049-9.00009-5
25. Sanderson, M. J., Smith, I., Parker, I., & Bootman, M. D. (2014). Fluorescence microscopy. *Cold Spring Harbal Protocols, 10*, 1042–1065. https://doi.org/10.1101/pdb.top071795
26. Nwaneshiudu, A., Kuschal, C., Sakamoto, F. H., Rox Anderson, R., Schwarzenberger, K., & Young, R. C. (2012). Introduction to confocal microscopy. *Journal of Investigative Dermatology, 132*(12), 1–5. https://doi.org/10.1038/jid.2012.429
27. Bharali, D. J., Lucey, D. W., Jayakumar, H., Pudavar, H. E., & Prasad, P. N. (2005). Folate-receptor mediated delivery of InP quantum dots for bioimaging using confocal and two-photon microscopy. *Journal of American Chemical Society, 127*(32), 11364–11371. https://doi.org/10.1021/ja051455x
28. Oleksiievets, N., et al. (2022). Single-molecule fluorescence lifetime imaging using wide-field and confocal-laser scanning microscopy: A comparative analysis. *Nano Letters, 22*(15), 6454–6461. https://doi.org/10.1021/acs.nanolett.2c01586
29. Low-Nam, S. T., et al. (2011). ErbB1 dimerization is promoted by domain co-confinement and stabilized by ligand binding. *Nature Structural & Molecular Biology, 18*(11), 1244–1249. https://doi.org/10.1038/nsmb.2135
30. Murar, M., Albertazzi, L., & Pujals, S. (2022). Advanced optical imaging-guided nanotheranostics toward personalized cancer drug delivery. *Nanomaterials, 12*, 399–416. https://doi.org/10.3390/nano12030399
31. Carlson, A. L., Gillenwater, A. M., Williams, M. D., El-Naggar, A. K., & Richards Kortum, R. R. (2007). Confocal microscopy and molecular-specific optical contrast agents for the detection of oral neoplasia. *Technology in Cancer Research and Treatment, 6*(5), 361–374. https://doi.org/10.1177/153303460700600501
32. Ishikawa-Ankerhold, H. C., Ankerhold, R., & Drummen, G. P. C. (2012). Advanced fluorescence microscopy techniques—FRAP, FLIP, FLAP, FRET and FLIM. *Molecules, 17*(4), 4047–4132. https://doi.org/10.3390/MOLECULES17044047
33. Clegg, R. M. (1995). Fluorescence resonance energy transfer. *Current Opinion in Biology, 6*, 103–113. https://doi.org/10.1016/0958-1669(95)80016-6
34. Wang, Y., & Geraghty, R. J. (2019). FRET-based assay using a three-way junction DNA substrate to identify inhibitors of human cytomegalovirus pUL89 endonuclease activity. *European Journal of Pharmaceutical Sciences, 127*, 29–37. https://doi.org/10.1016/j.ejps.2018.10.016

35. Il Joo, K., et al. (2008). Site-specific labeling of enveloped viruses with quantum dots for single virus tracking. *ACS Nano, 2*(8), 1553–1562. https://doi.org/10.1021/nn8002136
36. Dennis, A. M., Rhee, W. J., Sotto, D., Dublin, S. N., & Bao, G. (2012). Quantum dot-fluorescent protein fret probes for sensing intracellular pH. *ACS Nano, 6*(4), 2917–2924. https://doi.org/10.1021/nn2038077
37. Jakob, L. A., et al. (2021). Single photon multiclock lock-in detection by picosecond time-stamping. *Optica, 8*(12), 1646–1653. https://doi.org/10.1364/optica.441487
38. Hoebe, R. A., Van Oven, C. H., Gadella, T. W. J., Dhonukshe, P. B., Van Noorden, C. J. F., & Manders, E. M. M. (2007). Controlled light-exposure microscopy reduces photobleaching and phototoxicity in fluorescence live-cell imaging. *Nature Biotechnology, 25*(2), 249–253. https://doi.org/10.1038/nbt1278
39. Helmchen, F., & Denk, W. (2005). Deep tissue two-photon microscopy. *Nature Methods, 2*, 932–940. https://doi.org/10.1038/nmeth818
40. Li, Y., Guo, S., Mattison, B., Hu, J., Man, K. N. M., & Yang, W. microscopy with adaptive line-excitation. *Optica, 11*(8), 1138–1145. https://doi.org/10.1364/optica.529930
41. Mizuta, Y. (2021). Advances in two-photon imaging in plants. *Plant & Cell Physiology, 62*(8), 1224–1230. https://doi.org/10.1093/pcp/pcab062
42. Wu, P. C., et al. (2018). Efficient two-photon luminescence for cellular imaging using biocompatible nitrogen-doped graphene quantum dots conjugated with polymers. *Nanoscale, 10*(1), 109–117. https://doi.org/10.1039/c7nr06836k
43. Ohulchanskyy, T. Y., Roy, I., Yong, K. T., Pudavar, H. E., & Prasad, P. N. (2010). High-resolution light microscopy using luminescent nanoparticles. *WIREs Nanomedicine and Nanobiotechnology, 2*, 162–175. https://doi.org/10.1002/wnan.67
44. Zhang, N., et al. (2019). Cell imaging using two-photon excited CdS fluorescent quantum dots working within the biological window. *Nanomaterials, 9*(3), 369–381. https://doi.org/10.3390/nano9030369
45. Meddens, M. B. M., de Keijzer, S., & Cambi, A. (2014). High spatiotemporal bioimaging techniques to study the plasma membrane nanoscale organization. In *Fluorescence microscopy: Super-resolution and other novel techniques* (pp. 49–63). https://doi.org/10.1016/B978-0-12-409513-7.00004-X
46. Wüstner, D., Solanko, L. M., Lund, F. W., Sage, D., Schroll, H. J., & Lomholt, M. A. (2012). Quantitative fluorescence loss in photobleaching for analysis of protein transport and aggregation. *BMC Bioinformatics, 13*(1), 296–317. https://doi.org/10.1186/1471-2105-13-296
47. Hauth, J., Chodorski, J., Wirsen, A., & Ulber, R. (2020). Improved FRAP measurements on biofilms. *Biophysics Journal, 118*(10), 2354–2365. https://doi.org/10.1016/j.bpj.2020.03.017
48. Lorén, N., et al. (2015). Fluorescence recovery after photobleaching in material and life sciences: Putting theory into practice. *Quarterly Reviews of Biophysics, 48*(3), 323–387. https://doi.org/10.1017/S0033583515000013
49. Berezin, M. Y., & Achilefu, S. (2010). Fluorescence lifetime measurements and biological imaging. *Chemical Review, 110*(5), 2641–2684. https://doi.org/10.1021/cr900343z
50. Perri, A., et al. (2018). Time- and frequency-resolved fluorescence with a single TCSPC detector via a Fourier-transform approach. *Optica Express, 26*(3), 2270–2279. https://doi.org/10.1364/oe.26.002270
51. Huang, M., et al. (2020). Carbon dots for intracellular pH sensing with fluorescence lifetime imaging microscopy. *Nanomaterials, 10*, 604. https://doi.org/10.3390/nano10040604
52. Goreham, R. V., Schroeder, K. L., Holmes, A., Bradley, S. J., & Nann, T. (2018). Demonstration of the lack of cytotoxicity of unmodified and folic acid modified graphene oxide quantum dots, and their application to fluorescence lifetime imaging of HaCaT cells. *Microchimica Acta, 185*, 128. https://doi.org/10.1007/s00604-018-2679-8

Quantum Dots in the Biomedical Field

Kaushik Kokil Nath, Gazi Ameen Ahmed, and Rajib Biswas

1 Introduction

Quantum dots are semiconductor particles with nanometer-scale dimensions that exhibit unique optical and electronic properties attributed to their size and shape. These nanocrystals ranging between 2 and 10 nm in diameter, display quantum mechanical effects. Quantum dots are composed of semiconductor materials, such as cadmium selenide, cadmium sulfide, and indium phosphide. These materials have characteristic photoluminescent properties. In these nanostructures, electron and hole motion becomes discretized, resulting in the development of discrete energy levels, similar to atoms, with size-dependent emission spectra [1].

The concept of quantum dots was first described by Alexei Ekimov in the early 1980s, when he reported size-dependent quantum effects in glass matrices containing semiconductor nanocrystals [2]. Louis Brus further advanced this area of research further by showing the quantum size effects in colloidal semiconductor nanocrystals [3]. These early findings laid the groundwork for the synthesis and application of QDs in a wide variety of fields. In the 1990s, breakthroughs in colloidal synthesis techniques allowed high-quality QDs to be fabricated with much more controllable sizes and shapes [4]. A critical innovation during this decade was the development of optically efficient and stable core–shell QDs [5]. At the end of the 1990s, the development of water-soluble QDs gave researchers an improved understanding of their use in biological and biomedical applications [6]. By the

K. K. Nath · G. A. Ahmed · R. Biswas (✉)
Laboratory for Plasma Processing of Materials, Department of Physics, Tezpur University, Tezpur, India
e-mail: rajib@tezu.ernet.in

S. P. Kabekkodu et al. (eds.), *Functional Quantum Dots*, Lecture Notes in Nanoscale Science and Technology 40,
https://doi.org/10.1007/978-3-031-98219-4_3

2000s, QDs had already established positions in multiple diagnostic and imaging technologies, firmly putting them into the biomedical mainstream [7].

Quantum dots possess unique optical properties since most semiconducting materials are subjected to quantum confinement, wherein the size of a nanocrystal is nearer or smaller than the exciton Bohr radius of that material. This causes discrete energy levels with a size-tunable bandgap that can cause QDs to emit light at specific wavelengths when excited. By varying their size, one can obtain a QD that emits light across the visible spectrum, from blue to red. In summary, the emission color of QDs is highly dependent on their size: the smaller the QD is, the bluer the emitted light; the larger the QD is, the redder it is. The high photostability and broad absorption spectra of QDs make them very good candidates for a wide range of applications in optics [8]. Unlike organic dyes, which bleach easily; QDs are well suited for long-term imaging and tracking of biological systems. Moreover, their wide absorption spectra enable multiple QDs to be excited using only one light source, allowing multiplexed imaging and detection [9]. When the dimensions of a semiconductor nanocrystal are comparable to or smaller in magnitude than the exciton Bohr radius, charge carriers become confined, resulting in discrete energy levels. This confinement impacts the electronic structure of the material and produces size-dependent optical properties. Quantum confinement not only affects the wavelength of the light that is emitted but also enhances the fluorescence quantum yield and photostability of QDs [8]. These characteristics result in bright and stable fluorescence, which is one of the prerequisites for biomedical imaging and diagnostic applications.

2 Synthesis of Quantum Dots

2.1 Methods of Synthesis

2.1.1 Colloidal Synthesis

Colloidal synthesis is one of the most commonly used process in the preparation of quantum dots (QDs). Colloidal synthesis involves the nucleation and growth of nanocrystals in a liquid medium, typically using organometallic precursors and coordinating solvents. By precisely controlling reaction variables such as temperature, concentration, and reaction time, this method can easily yield QDs that are uniform in both size and shape. The advantages of colloidal synthesis include scalability, reproducibility, and high-quality QDs with tunable optical properties. It has been optimized to produce monodisperse QDs with high quantum yields. In addition, surface modification techniques can be readily integrated with colloidal synthesis to increase the stability and biocompatibility of QDs [10].

2.1.2 Vapor-Phase Synthesis

Vapor-phase techniques, collectively known as chemical vapor deposition (CVD), are alternative methods for synthesizing quantum dots. In such techniques, precursor gases are introduced into a reactor and undergo chemical reactions to form nanocrystals on a substrate. Vapor-phase synthesis offers precise control over the composition, size, and morphology of QDs [11]. A key advantage of vapor-phase synthesis is for concision and emphasis is the high purity and crystallinity of the QDs produced. However, it usually requires specialized equipment and is relatively less scalable than colloidal synthesis. Vapor-phase synthesis is used in making QDs, which is intended for optoelectronic applications where very high quality and purity are needed.

2.1.3 Hydrothermal/Solvothermal Synthesis

Hydrothermal and solvothermal synthesis methods involve nanocrystal growth in high-temperature and high-pressure aqueous or nonaqueous solutions. Specifically, these methods are particularly suited for preparing QDs with complex compositions and structures, such as core–shell and alloyed QDs. The hydrothermal/solvothermal synthesis provides advantage in precise control of the size and shape of QDs by adjusting the reaction conditions. In addition, it is environmentally friendly, and the obtained QDs are highly crystalline and stable [12]. However, the processes can be time-consuming and may require specialized equipment.

2.2 *Surface Modification and Functionalization*

2.2.1 Ligand Exchange

Surface ligands play crucial roles in determining the stability, solubility, and biocompatibility of quantum dots. Ligand exchange is a common technique used to replace the native hydrophobic ligands on QDs with hydrophilic ligands, making them water-soluble and suitable for biological applications. This process involves incubating QDs with new ligands that have higher binding affinities for the QD surface. The choice of ligands for exchange can significantly impact the optical properties and stability of QDs. Common ligands used for biocompatible QDs include thiol-containing molecules, polymers, and peptides. Ligand exchange not only enhances the solubility of QDs in biological buffers but also provides functional groups for further bioconjugation [13].

2.2.2 Polymer Coating

One surface engineering technique that has been implemented to increase quantum dot stability and biocompatibility is the use of a polymer coating [28]. These polymers, which are mainly used to coat QDs, include polyethylene glycol, polyvinyl alcohol, and polydopamine [29]. These polymers provide a protective layer that prevents aggregation and improves dispersion in an aqueous environment. Polymer-coated QDs exhibit reduced toxicity and extended circulation time in biological systems, polymer-coated QDs have the potential to serve as excellent platforms for in vivo applications. More importantly, a polymer coat can be functionalized with targeting ligands, drugs, or imaging agents to create a multifunctional QD-based platform for diagnostics and therapeutics [14].

2.2.3 Bioconjugation Techniques

Bioconjugation is a process that involves antibodies, peptides, or nucleic acids are attached to a QD surface. Functionalization enables QDs to bind only with target biomolecules, cells, or tissues to enable targeted imaging, diagnostics, and therapy. Classic examples of bioconjugation methods for QDs include covalent coupling, electrostatic, and affinity-based interactions. Covalent coupling methods, such as EDC/NHS chemistry, produce stable and specific attachments of biomolecules to QDs. Electrostatic and affinity-based interactions, on the other hand, afford reversible and tunable conjugation, which may be dynamically regulated to control the functionality of QDs [15].

3 Characterization of Quantum Dots

3.1 Structural Characterization

3.1.1 Transmission Electron Microscopy (TEM)

Transmission electron microscopy (TEM) is a powerful technique used to characterize the size, shape, and crystal structure of quantum dots. TEM provides high-resolution images by transmitting a beam of electrons through a thin sample, allowing for direct observation of individual QDs. TEM analysis can reveal the size distribution and morphology of QDs, as well as their internal crystal structure. High-resolution TEM (HRTEM) can provides detailed insights into the lattice fringes and defects within QDs, which are critical essential for understanding their optical and electronic properties [16]. TEM is often complemented by techniques, such as scanning electron microscopy (SEM) and atomic force microscopy (AFM), to provide a comprehensive characterization of QDs.

3.1.2 X-Ray Diffraction (XRD)

X-ray diffraction (XRD) is a technique used to determine the crystalline structure and phase composition of quantum dots. XRD involves directing X-rays onto a sample and measuring the diffraction pattern produced by the interaction of the X-rays with the crystal lattice. The diffraction pattern provides information on the lattice parameters, crystallite size, and phase purity of the QDs. XRD is particularly useful for identifying the crystal structure of QDs and confirming the presence of specific phases or impurities [17]. This technique is essential for optimizing synthesis conditions and ensuring the quality of QDs for biomedical applications.

3.2 Optical Characterization

3.2.1 Absorption and Photoluminescence Spectroscopy

Absorption and photoluminescence (PL) spectroscopy are fundamental techniques used to characterize the optical properties of quantum dots (QDs). Absorption spectroscopy measures the absorption of light by QDs at different wavelengths, providing information on their bandgap and electronic transitions [18]. Photoluminescence spectroscopy involves exciting QDs with a light source and measuring the emitted fluorescence. The emission spectrum provides information on the size distribution, quantum yield, and photostability of QDs [19]. Time-resolved PL spectroscopy further elucidates the dynamics of exciton recombination and the presence of surface states or defects [20].

3.2.2 Time-Resolved Fluorescence

Time-resolved fluorescence spectroscopy measures the decay of fluorescence over time, providing insights into the excited-state dynamics and lifetimes of quantum dots. This technique involves exciting QDs with a short light pulse and monitoring the fluorescence decay using a fast detector [21]. Time-resolved fluorescence is used to study the recombination kinetics of excitons and the influence of surface states or traps on the emission properties of QDs [22]. This information is crucial for optimizing the design and functionalization of QDs for biomedical applications, where long fluorescence lifetimes and high quantum yields are desirable.

3.3 *Surface Characterization*

3.3.1 Fourier Transform Infrared Spectroscopy (FTIR)

Fourier transform infrared (FTIR) spectroscopy is used to analyze the surface chemistry and functional groups present on the quantum dots. FTIR measures the absorption of infrared light by molecular vibrations, providing a fingerprint of the chemical bonds and functional groups on the QD surface [23]. FTIR is particularly useful for characterizing surface modifications and the functionalization of QDs. By comparing the FTIR spectra before and after surface modification, one can identify the presence of specific ligands, polymers, or biomolecules attached to the QD surface [24]. This information is essential for ensuring the stability and functionality of QDs in biological environments.

3.3.2 X-Ray Photoelectron Spectroscopy (XPS)

X-ray photoelectron spectroscopy (XPS) is a surface-sensitive technique used to analyze the elemental composition and chemical states of atoms on the surface of quantum dots. XPS involves irradiating a sample with X-rays and measuring the kinetic energy of the emitted photoelectrons. XPS provides quantitative information on the surface composition and oxidation states of elements in QDs [25]. This technique is used to confirm the presence of specific surface ligands, assess the degree of oxidation, and evaluate the effectiveness of surface modifications [26]. XPS is essential for understanding the interactions between QDs and their biological environment, as well as for optimizing surface functionalization strategies.

4 Quantum Dots in Diagnostics

4.1 *Imaging Applications*

4.1.1 Fluorescence Imaging

Among the more popular applications of quantum dots in the field of biomedicine is fluorescence imaging. Due to their unique optical properties of QDs, including size-tunable emission and high photostability, they are perfect fluorophores for imaging biological samples. QDs can be linked to targeting molecules, such as antibodies or peptides, which specifically label cells, tissues, or biomolecules. Quantum dots have been further combined into many fluorescence imaging techniques, including confocal microscopy, total internal reflection fluorescence microscopy, and even single-molecule imaging. These techniques enable high-resolution visualization of structures, dynamic processes, and molecular interactions in a physiologically relevant environment with very high resolution. Since the multiplexing of QDs with many emission wavelengths is possible, their value was enhanced for complex imaging approaches [27].

4.1.2 Multimodal Imaging

From the name, multimodal imaging is an imaging method in which two or more imaging modalities are used for the provision of information with respect to biological systems. Quantum dots can be integrated with other imaging agents, such as magnetic nanoparticles or contrast agents, that form multifunctional probes to allow for multimodal imaging. As a result, functional QDs can be combined with magnetic resonance imaging, which employs high spatial resolution together with the high sensitivity and specificity of QD fluorescence imaging. Therefore, these conjugates allow for anatomical and molecular imaging simultaneously, offering full details of biological processes. Multimodal imaging using QDs has been applied in cancer diagnosis, tracking of stem cells, and monitoring the progression of diseases [28].

4.2 *Biosensing Applications*

4.2.1 Quantum Dot-Based Biosensors

Quantum dot-based biosensors have been developed on the working principle of using the unique optical properties of QDs for detecting particular biomolecules, pathogens, or environmental pollutants. Typically, these biosensors involve the conjugation of QDs with recognition elements such as antibodies, aptamers, or enzymes that bind to target analyte [29]. The interaction of QD-conjugated recognition elements with the target analyte changes the QDs' optical properties—fluorescence intensity, wavelength shift, or quantitatively measurable Förster resonance energy transfer [30]. These changes are then detected quantitatively to determine the presence and concentration of the target analyte. QD-based sensors offer high selectivity, sensitivity, and multiplexing capabilities, enhancing biosensing applications in biomedical diagnostics and food safety to correct the misuse of inoculated [29].

4.2.2 Detection of Biomarkers and Pathogens

Quantum dots have been applied to develop a series of biosensors for different biomarkers and pathogens that result in diseases. For example, QD-based immunoassays enable the detection of cancer biomarkers such as prostate-specific antigen and carcinoembryonic antigen with very high sensitivity and specificity [31]. On the other hand, QD-based nucleic acid sensors can detect viral and bacterial DNA or RNA, thus enabling fast and accurate infectious disease diagnosis. The high photostability and multiple detection capabilities of QDs allow for the simultaneous detection of multiple biomarkers or pathogens in one assay [32]. This multiplexed detection can be mostly applied in complex diseases, such as cancer and infectious diseases, that involve multiple biomarkers. QD-based biosensors have demonstrated great promise for point-of-care diagnostics, which enables both rapid and portable testing of some infectious diseases at the clinical and field levels.

4.3 In Vitro and In Vivo Diagnostics

4.3.1 Cell Labelling and Tracking

Due to their bright fluorescence and excellent photostability, quantum dots (QDs) are widely used for cell labeling, both in vitro and in vivo. Conversely, QD labelling of cell types can result from conjugation with related cell-specific ligands, such as antibodies or peptides. This type of labelling investigates cell processes such as proliferation, differentiation, and migration via direct visualization [33]. Labelling of cells with QDs via in vitro methods has become a standardized process for observing cellular behaviors and interactions either under a fluorescence microscope or while crossing through a flow cytometer [34]. In vivo cell tracking using QDs can aid in subsequent cell engraftment, localization, and migration inside a living subject. This technique has immense relevance in studies using stem cells, in cancer metastasis research, and in the study of the behavior of immune cells, among other processes [35].

4.3.2 In Vivo Imaging and Diagnostics

The use of quantum dots for in vivo imaging revealed real-time visualization of biological processes and the course of disease progression in living organisms. QDs can act as fluorescent probes for imaging tumors, vasculature, and other tissues in animal models. Their high photostability and brightness support long-term imaging, allowing monitoring of disease progression and treatment response over time [36]. Quantum dots have become a tool for most in vivo imaging modalities—such as near-infrared fluorescence imaging, photoacoustic imaging, and fluorescence lifetime imaging—to provide highly complementary information about the functional, molecular, and anatomical status of biological systems [27]. QD-based in vivo imaging is highly valuable for early diagnosis, surgical guidance, and drug efficiency monitoring in clinical practice.

5 Quantum Dots in Therapeutics

5.1 Drug Delivery Systems

5.1.1 Quantum Dot–Drug Conjugates

Quantum dots may be conjugated to therapeutic agents to formulate multifunctional drug delivery systems. QD–drug conjugates offer several advantages, including targeted delivery, controlled release, and real-time monitoring for drug distribution and efficacy [37]. QDs can be functionalized with targeting ligands such as antibodies or peptides to selectively deliver drugs to specified cells or tissues. The conjugation of QDs to drugs can improve their solubility, stability, and bioavailability [38]. The fluorescence properties of the QDs also allow for tracing drug delivery and release processes in real time, which will be very useful in studying the pharmacokinetics and biodistribution of therapeutic agents [39]. These conjugates show promise as therapeutics for cancer, infectious diseases, and other disorders.

5.1.2 Targeted Delivery Mechanisms

The use of targeted quantum dots for drug delivery involves specific ligands that cause the QD–drug conjugates to concentrate at the site of action. This can be achieved through passive, active, or stimuli-responsive targeting. Passive targeting occurs via the EPR effect, through which the QD–drug conjugate accumulates within tumor tissues because of its leaky vasculature. It utilizes the concept of specific ligands that bind to receptors or antigens on the surface of target cells [40]. This method enhances selectivity and efficiency for drug delivery, which reduces off-target effects and minimizes systemic toxicity. Controlled release is ensured through stimuli-responsive mechanisms, such as pH-sensitive or enzyme-responsive linkers, in response to certain physiological conditions [41].

5.2 Photodynamic Therapy (PDT)

Photodynamic therapy is a minimally invasive treatment that uses light-activated photosensitizers to generate reactive oxygen species (ROS), inducing cell death. Quantum dots play a dual role in PDT as photosensitizers or energy donors, enhancing the efficacy and specificity of the therapy. Upon photoexcitation, QDs transfer energy to molecular oxygen and form ROS, resulting in oxidative damage to cellular components [42].

Recent improvements in QD-based PDT have been directed toward increasing the targetability and selectivity of treatment for improved therapeutic efficacy. QDs can be conjugated to targeting ligands, to selectively accumulate in tumor tissues, minimizing damage to healthy tissues [5, 43]. Moreover, near-infrared QDs may enable deeper tissue penetration with reduced photodamage to the skin. Preclinical investigations of QD-based PDT have shown its potential for treating skin, breast, and prostate cancers [44]. The combination of QD-based PDT with other therapeutic modalities further enhances the therapeutic potential of this treatment approach, either by chemotherapy or immunotherapy [45]. Current research on optimizing the design and functionalization of QDs aims towards the clinical translation of PDT.

5.3 Gene Therapy

5.3.1 Quantum Dots as Gene Delivery Vectors

QDs can also function as gene delivery vectors for the introduction of therapeutic genes into target cells. The functionalization of the QD surface with nucleic acids such as plasmid DNA or siRNA makes them useful for applications in gene therapy. A large surface area, along with the multivalency of QDs, facilitates efficient loading and delivery of genetic material [46].

5.3.2 Applications in Genetic Diseases and Cancer Therapy

QD gene delivery has the potential to treat genetically induced diseases, such as cystic fibrosis and muscular dystrophy, and for applications in cancer therapy [47]. Another very important application of QDs is in the development of a gene delivery system by which therapeutic genes can be delivered to targeted diseased cells, minimizing off-target effects and hence developing efficient gene therapy [48]. The ability to trace QD-mediated gene delivery in real time explains the efficiency of transfection and therapeutic outcomes.

6 Toxicity and Biocompatibility

6.1 Cytotoxicity Studies

6.1.1 In Vitro Cytotoxicity Assays

Cytotoxicity studies are critical for establishing the safety and biocompatibility of quantum dots. In general, in vitro cytotoxicity assays usually estimate changes in viability for establishing safe and biocompatible quantum dots. In general, in vitro cytotoxicity assays usually estimate changes in viability, proliferation, and morphology following the culture of cells with QDs. The most often employed assays include MTT, LDH, and live/dead staining.

6.1.2 Dose–response relationships

Dose–response studies have revealed the dose at which QDs exert their cytotoxic effects. These studies have established safe and effective dosages that are considered safe and effective for application in the biomedical field. The physiochemical parameters affecting cytotoxicity include the QD composition, size, surface charge, and functionalization of the surface [49].

6.2 *In Vivo Toxicity*

6.2.1 Animal Model Studies

In vivo toxicity studies involve direct administration of QDs in animal models and time-to-time observations of their health status, behavior, organ functioning, etc. These studies are necessary to explain the biodistribution, metabolism, and long-term effects of QDs. Histopathological analysis of tissues has revealed potential toxic effects at the cellular level [50]. QDs are semiconductor nanoparticles with special optical and electronic properties that assist in biomedical applications such as imaging, diagnostics, and drug delivery. However, clinical use has faced concerns related to the potential long-term toxicity and biodistribution of QDs.

6.2.2 Accumulation with Tissue

QDs persist in organs such as the liver, spleen, and kidneys, eliciting chronic toxicity. This accumulation can lead to reactive oxygen species (ROS) generation, inflammation, and cellular damage, especially with QDs based on heavy metals such as cadmium [51].

6.2.3 Oxidative Stress and Inflammation

QDs generate reactive oxygen species, which cause oxidative stress that can damage DNA, proteins, and lipids [9]. This damage may trigger chronic inflammation—a condition implicated in many diseases, including fibrosis and cancer [51].

6.2.4 Genotoxicity and Carcinogenicity

Quantum dots (QDs) may induce DNA damage, raising concerns about their genotoxic and carcinogenic potential. The interaction between QDs and genetic materials could result in DNA strand breaks, mutations, and chromosomal aberrations; thus, long-term studies that can help explain and mitigate these risks are warranted [52].

6.2.5 Biodistribution

6.2.5.1 Organ-Specific Distribution

Typically, QDs are taken up by the reticuloendothelial system (RES) and accumulate in the liver, spleen, and lymph nodes. QDs smaller than 10 nm in size are more likely to be cleared by renal filtration, and larger QDs are sequestered in the RES [53].

6.2.6 Clearance Mechanisms

Quantum dots (QDs) are cleared through renal and hepatic pathways. Smaller and more hydrophilic QDs are renally cleared, whereas larger and more hydrophobic QDs undergo hepatic and bile clearance [54]. Surface engineering with biocompatible polymers such as PEG increases their circulation time and hence enhances their clearance [55].

6.2.7 Challenges in Long-Term Retention and Clearance

However, long-term retention-associated clinical challenges associated with QDs remain to be overcome. One such way to increase clearance in this regard involves the use of biodegradable QDs, which metabolize or are excreted as nontoxic byproducts. The engineering of QDs for rapid renal and hepatic clearance can reduce their long-term toxicity [56].

6.3 Strategies to Mitigate Toxicity

QDs are semiconducting nanoparticles with special optical and electronic properties that allow them to be health targeted for imaging, diagnosis, and therapeutics in biomedicine. However, such applications are sharply limited by their innate toxicity. Several strategies have been developed to enhance their safety and increase the safety of these materials for clinical applications.

6.3.1 Surface Modification

Surface Coating: The surface coating of QDs uses biocompatible materials, such as polyethylene glycol, silica, and phospholipid micelles. This allows the core to prevent the release of toxic ions in the body and protection from toxic biological interactions. Accordingly, the stability and circulation time of the QDs are increased. The process of functionalization of quantum dots includes attachment to targeting ligands, such as antibodies, peptides, and small molecules, to increase their specificity to target cells, thereby resulting in a decrease in off-target effects, as well as the attenuation of toxicity that a quantum dot usually possesses, thus completing its place of accumulation [57].

6.3.2 Biodegradable Quantum Dot Composition

QDs made up of metabolically or renally cleared materials, e.g., zinc oxide (ZnO) or iron oxide (Fe_2O_3), decompose into nontoxic byproducts, avoiding long-term toxicity associated with degradation products [58, 59]. Self-decomposing QDs: Similarly, QDs that decompose in vivo into nontoxic building blocks under physiologically relevant conditions avoid the accumulation of QDs in tissues, allowing safer long-term in vivo applications [60].

6.3.3 Doping and Alloy Doping

When small amounts of less toxic metals are doped with a Cd-based QD structure, the toxic thallium group can be largely mitigated by simply changing the excitonic and electronic nature of the NPs and reducing the release of toxic ions [61]. However, alloyed QDs with a mixture of components due to several semiconductor components have less toxicity because of the solidifying effect of the compounded materials [62].

6.3.4 Size and Shape Optimization Size Control

Smaller QDs (<5 nm) are easily eliminated via renal filtration, minimizing long-term toxicity. By tailoring the size to find the functionality–clearance balance, toxic effects are mitigated. Changing the shape of a QD, as seen in the development of rod-shaped or elongated QDs, changes its cellular uptake; these changes further influence its biocompatibility and toxic interactions [63].

6.3.5 Heavy Metal Content Reduction

Alternative materials: QDs prepared from alternative, less toxic materials, such as carbon dots, silicon QDs, or graphene quantum dots, should be less toxic [64]. Heavy metal-free QDs: QDs composed of nontoxic elements, such as indium phosphide (InP) or copper indium sulfide (CIS), also exhibit the effective optical properties of conventional QDs without causing toxicity concerns [65].

6.3.6 In Vivo Clearance Mechanisms Passive Targeting

QDs are engineered in a manner that results in minimal physical trapping. They remain predominantly in circulation, with less accumulation in off-target organs. They are designed for active excretion via renal or hepatic pathways minimizes accumulation in the body and, in turn, reduce long-term toxicity. This active targeting and clearance may be significantly increased by designing recognition sites for scavenger receptors on macrophages, which eventually minimize QD-induced toxicity [66]. Collectively, these strategies enhance the safety profile of QDs and allow broader usage.

6.3.7 Regulatory and Ethical Considerations

For biomedical applications embody guidelines from agencies such as the FDA, EMA, and ISO. Compliance with regulatory requirements provides safe and effective use of QDs in clinical settings. Case studies of approved QD-based products illustrate regulatory pathways and challenges for the translation of QD-based technologies to the market. These case studies further illuminate the clinical translation and commercialization of QD-based diagnostics and therapeutics.

Such clinical application of quantum dots, therefore, brings with it concerns associated with informed consent, safety toward patients, and equable access to QD-based technologies. Ethical guidelines provide the general public with assurance that the benefits arising from QD-based applications outweigh the risks and that patients are properly informed about possible risks and benefits. Public perception and acceptance of quantum dot-based technologies will be necessary; hence, their adoption and use in biomedical applications will be affected. This means that there must be openness in communication, public engagement, and education on QD safety and the benefits that will accrue to allay fears and win public acceptance.

7 Future Prospects and Challenges

The engineering of QD-based nanoconjugates by their encapsulation into other kinds of nanotechnology, such as nanofibers [67] and hydrogels [68], has driven their biomedical applications toward multifunctional purposes. These hybrid systems combine QD optical and electronic properties with the structural and functional attributes of other nanomaterials. Advances in the synthesis, surface modification, and functionalization of quantum dots continue to lead to new biomedical applications. Indeed, as innovations in QD technology proceed, they have focused on improving their performance while reducing their toxicity and extending their applications in diagnosis, therapeutics, and imaging [69]. This clearly suggests that, clinically, one of the major translational challenges facing quantum dots remains how to determine their cellular uptake, biological immune response, and clearance. Surface modification, targeting ligands, and biodegradable QDs are some of the strategies put in place to help combat these challenges [70]. Prerequisites for the large-scale application of QDs in biomedicine include both high-quality quantum dots and scalability. The methods for synthesis and quality control together with manufacturing procedures will further need to mature when QDs are expected to become stable in the clinic and industry [71].

Quantum dots in this direction can be game changers for healthcare revolutions by providing precision medicine, in vitro diagnostics, and targeted treatment. From distinct optical and electronic features, an avenue for developing personal diagnostic tools and therapies that may fit the needs of each individual patient is possible [72].

8 Conclusion

This chapter provides an overview of the distinctive properties, synthesis, characterization, and biomedical applications of quantum dots. Quantum dots are semiconductor nanoparticles with unique optical and electronic properties arising from quantum confinement, which occurs when their dimensions approach the exciton Bohr radius, resulting in size-dependent emission spectra. These properties make QDs ideal candidates for optical applications, such as long-term imaging in biological systems and subsequent tracking of movements therein. The routes generated for the synthesis of QDs include physical, colloidal, vapor-phase, and hydrothermal/solvothermal routes. Each of these methods has advantages and limitations. The focusing frontier involves the development of surface modification strategies for QDs, such as ligand exchange, polymer coating, and the formation of bioconjugates, to enhance QD stability, biocompatibility, and functionality for biomedical applications. Characterization will be carried out via techniques such as TEM, XRD, absorption and photoluminescence spectroscopy, time-resolved fluorescence, FTIR, and XPS to determine their structural, optical, and surface properties in detail. These QDs take part in fluorescence imaging, multimodal imaging, and biosensing associated with diagnostic applications, allowing high-resolution imaging and multiplex detection of biomolecules, pathogens, and cellular processes. For example, in therapeutics, QDs are employed as drug delivery systems, photosensitizers in photodynamic therapy, and gene delivery vectors. However, with the full and proper execution of in vitro and in vivo studies, their cytotoxicity and long-term toxicity cannot be overlooked, and mitigated by surface modifications and biodegradable QD designs. The regulatory frameworks and ethical considerations available are related to the safe and effective clinical application of QDs. However, the trends that are developing—the integration of QDs with other nanotechnologies and advances in QD technology itself—are further increasing their potential in the biomedical field, notwithstanding the existing biological barrier challenges, scalability, and commercial production. The future of QDs in personalized medicine is promising, offering innovative solutions for precision diagnostics and targeted treatments, ultimately leading to the emergence of new paradigms in nanomedicine.

Acknowledgement The authors would like to acknowledge Tezpur University, Assam, India.

References

1. Sumanth Kumar, D.; Jai Kumar, B.; Mahesh, H.M. Chapter 3—Quantum nanostructures (QDs) overview. Available online: https://www.sciencedirect.com/science/article/abs/pii/B9780081019757000038. Accessed on 1 August 2024.
2. Manna, L. (2023). The bright and enlightening science of quantum dots. *Nano Letters, 23*, 9673–9676. https://doi.org/10.1021/acs.nanolett.3c03904
3. Peng, X., Manna, L., Yang, W., Wickham, J., Scher, E., Kadavanich, A., & Alivisatos, A. P. (2000). Shape control of CdSe nanocrystals. *Nature, 404*, 59–61. https://doi.org/10.1038/35003535
4. Dabbousi, B. O., Rodriguez-Viejo, J., Mikulec, F. V., Heine, J. R., Mattoussi, H., Ober, R., Jensen, K. F., & Bawendi, M. G. (1997). (CdSe)ZnS Core−Shell quantum dots: Synthesis and characterization of a size series of highly luminescent Nanocrystallites. *The Journal of Physical Chemistry. B, 101*, 9463–9475. https://doi.org/10.1021/jp971091y
5. Chan, W. C. (1998). Quantum dot bioconjugates for ultrasensitive nonisotopic detection. *Science, 281*, 2016–2018. https://doi.org/10.1126/science.281.5385.2016
6. Michalet, X. (2005). Quantum dots for live cells, in vivo imaging, and diagnostics. *Science, 307*, 538–544. https://doi.org/10.1126/science.1104274
7. Brus, L.E. Electron–Electron and Electron-Hole Interactions in Small Semiconductor Crystallites: The Size Dependence of the Lowest Excited Electronic State. The Journal of Chemical Physics 1984, 80, 4403–4409, doi:https://doi.org/10.1063/1.447218.
8. Biju, V., Itoh, T., Anas, A., Sujith, A., & Ishikawa, M. (2008). Semiconductor quantum dots and metal nanoparticles: Syntheses, optical properties, and biological applications. *Analytical and Bioanalytical Chemistry, 391*, 2469–2495. https://doi.org/10.1007/s00216-008-2185-7
9. Gao, X., Cui, Y., Levenson, R. M., Chung, L. W. K., & Nie, S. (2004). In vivo cancer targeting and imaging with semiconductor quantum dots. *Nature Biotechnology, 22*, 969–976. https://doi.org/10.1038/nbt994
10. Pu, Y., Cai, F., Wang, D., Wang, J.-X., & Chen, J.-F. (2018). Colloidal synthesis of semiconductor quantum dots toward large-scale production: A review. *Industrial and Engineering Chemistry Research, 57*, 1790–1802. https://doi.org/10.1021/acs.iecr.7b04836
11. Bera, D., Qian, L., Tseng, T.-K., & Holloway, P. H. (2010). Quantum dots and their multimodal applications: A review. *Materials, 3*, 2260–2345. https://doi.org/10.3390/ma3042260
12. Xu, M., Li, Z., Zhu, X., Hu, N., Wei, H., Yang, Z., & Zhang, Y. (2013). Hydrothermal/Solvothermal synthesis of graphene quantum dots and their biological applications. *Nano Biomedical Engineering, 5*. https://doi.org/10.5101/nbe.v4i3.p65-71
13. Yan, F., Jiang, Y., Sun, X., Bai, Z., Zhang, Y., & Zhou, X. (2018). Surface modification and chemical functionalization of carbon dots: A review. *Microchimica Acta, 185*. https://doi.org/10.1007/s00604-018-2953-9
14. Hezinger, A. F. E., Teßmar, J., & Göpferich, A. (2008). Polymer coating of quantum dots—A powerful tool toward diagnostics and sensorics. *European Journal of Pharmaceutics and Biopharmaceutics, 68*, 138–152. https://doi.org/10.1016/j.ejpb.2007.05.013
15. Foubert, A., Beloglazova, N. V., Rajkovic, A., Sas, B., Madder, A., Goryacheva, I. Y., & De Saeger, S. (2016). Bioconjugation of quantum dots: Review & impact on future application. *TrAC Trends in Analytical Chemistry, 83*, 31–48. https://doi.org/10.1016/j.trac.2016.07.008

16. Liu, C., Miller, P., Henstrom, W. L., & Gibson, J. (2000). Transmission electron microscopy of semiconductor quantum dots. *Journal of Microscopy, 199*, 130–140. https://doi.org/10.1046/j.1365-2818.2000.00729.x
17. Calvin, J. J., Kaufman, T. M., Sedlak, A. B., Crook, M. F., & Alivisatos, A. P. (2021). Observation of ordered organic capping ligands on semiconducting quantum dots via powder X-ray diffraction. *Nature Communications, 12*, 2663. https://doi.org/10.1038/s41467-021-22947-x
18. Seidl, S., Kroner, M., Dalgarno, P. A., Högele, A., Smith, J. M., Ediger, M., Gerardot, B. D., Garcia, J. M., Petroff, P. M., Karrai, K., et al. (2005). Absorption and photoluminescence spectroscopy on a single self-assembled charge-tunable quantum dot. *Physical Review B, 72*. https://doi.org/10.1103/physrevb.72.195339
19. Valenta, J., Juhasz, R., & Linnros, J. (2002). Photoluminescence spectroscopy of single silicon quantum dots. *Applied Physics Letters, 80*, 1070–1072. https://doi.org/10.1063/1.1448400
20. Koole, R., Schapotschnikow, P., Thijs, T. J. H., Vlugt, C., & Meijerink, A. (2008). Time-dependent photoluminescence spectroscopy as a tool to measure the ligand exchange kinetics on a quantum dot surface. *ACS Nano, 2*, 1703–1714. https://doi.org/10.1021/nn8003247
21. Wang, X., Wu, X., Lu, Z., & Tao, X. (2020). Comparative study of time-resolved fluorescent nanobeads, quantum dot nanobeads and quantum dots as labels in fluorescence immunochromatography for detection of aflatoxin B1 in grains. *Biomolecules, 10*, 575. https://doi.org/10.3390/biom10040575
22. Woodall, D. L., Tobias, A. K., & Jones, M. (2016). Resolving carrier recombination in CdS quantum dots: A time-resolved fluorescence study. *Chemical Physics, 471*, 2–10. https://doi.org/10.1016/j.chemphys.2015.10.011
23. Guerrero-Pérez, M. O., & Patience, G. S. (2019). Experimental methods in chemical engineering: Fourier Transform Infrared Spectroscopy—FTIR. *Canadian Journal of Chemical Engineering, 98*, 25–33. https://doi.org/10.1002/cjce.23664
24. Santos, J., Mansur, A., & Mansur, H. (2013). One-step biofunctionalization of quantum dots with chitosan and N-Palmitoyl chitosan for potential biomedical applications. *Molecules, 18*, 6550–6572. https://doi.org/10.3390/molecules18066550
25. Babu, B., Neelakanta Reddy, I., Yoo, K., Kim, D., & Shim, J. (2018). Bandgap tuning and XPS study of SnO2 quantum dots. *Materials Letters, 221*, 211–215. https://doi.org/10.1016/j.matlet.2018.03.107
26. Silva, F. O., Carvalho, M. S., Mendonça, R., Macedo, W. A., Balzuweit, K., Reiss, P., & Schiavon, M. A. (2012). Effect of surface ligands on the optical properties of aqueous soluble CdTe quantum dots. *Nanoscale Research Letters, 7*, 536. https://doi.org/10.1186/1556-276x-7-536
27. Wegner, K. D., & Hildebrandt, N. (2015). Quantum dots: Bright and versatile in vitro and in vivo fluorescence imaging biosensors. *Chemical Society Reviews, 44*, 4792–4834. https://doi.org/10.1039/c4cs00532e
28. Koole, R., Mulder, W. J. M., van Schooneveld, M. M., Strijkers, G. J., Meijerink, A., & Nicolay, K. (2009). Magnetic quantum dots for multimodal imaging. *Wiley Interdisciplinary Reviews: Nanomedicine and Nanobiotechnology, 1*, 475–491. https://doi.org/10.1002/wnan.14
29. Zhang, Y.; Cai, N.; Chan, V. Recent advances in silicon quantum dot-based fluorescent biosensors. Biosensors 2023, 13, 311–311, doi:doi:https://doi.org/10.3390/bios13030311.
30. Chou, K., & Dennis, A. (2015). Förster resonance energy transfer between quantum dot donors and quantum dot acceptors. *Sensors, 15*, 13288–13325. https://doi.org/10.3390/s150613288
31. Ding, L., Chen, X., He, L., Yu, F., Yu, S., Wang, J., Tian, Y., Wang, Y., Wu, Y., Liu, L.-E., et al. (2020). Fluorometric immunoassay for the simultaneous determination of the tumor markers carcinoembryonic antigen and cytokeratin 19 fragment using two kinds of CdSe/ZnS quantum dot nanobeads and magnetic beads. *Mikrochimica Acta, 187*, 171. https://doi.org/10.1007/s00604-019-3914-7
32. Liu, Y., Zhang, H., & Xu, Y. (2022). Quantum dots for pathogenic bacterial monitoring and combating. *Advanced Optical Materials, 11*. https://doi.org/10.1002/adom.202201826

33. Liu, J.; Cao, L.; LeCroy, G.E.; Wang, P.; Meziani, M.J.; Dong, Y.; Liu, Y.; Luo, P.G.; Sun, Y. Carbon "Quantum" Dots for Fluorescence Labelling of Cells. ACS Applied Materials & Interfaces 2015, 7, 19439–19445, doi:doi:https://doi.org/10.1021/acsami.5b05665.
34. Parak, W. J., Pellegrino, T., & Plank, C. (2005). Labelling of cells with quantum dots. *Nanotechnology, 16*, R9–R25. https://doi.org/10.1088/0957-4484/16/2/r01
35. Zhou, J., Zhao, R., Du, Y., Liu, S., Li, W., Gai, S., He, F., Feng, L., & Yang, P. (2022). A Si-CdTe composite quantum dots probe with dual-wavelength emission for sensitively monitoring intracellular H2O2. *Advanced Functional Materials, 32*. https://doi.org/10.1002/adfm.202112083
36. Xing, Y., & Rao, J. (2008). Quantum dot bioconjugates for in vitro diagnostics & in vivo imaging. *Cancer Biomarkers, 4*, 307–319. https://doi.org/10.3233/cbm-2008-4603
37. Soumya, K., More, N., Choppadandi, M., Aishwarya, D. A., Singh, G., & Kapusetti, G. (2023). A comprehensive review on carbon quantum dots as an effective photosensitizer and drug delivery system for cancer treatment. *Biomedical Technology, 4*, 11–20. https://doi.org/10.1016/j.bmt.2023.01.005
38. Jijie, R., Barras, A., Bouckaert, J., Dumitrascu, N., Szunerits, S., & Boukherroub, R. (2018). Enhanced antibacterial activity of carbon dots functionalized with ampicillin combined with visible light triggered photodynamic effects. *Colloids and Surfaces B: Biointerfaces, 170*, 347–354. https://doi.org/10.1016/j.colsurfb.2018.06.040
39. Ghaderi, S., Ramesh, B., & Seifalian, A. M. (2010). Fluorescence nanoparticles "quantum dots" as drug delivery system and their toxicity: A review. *Journal of Drug Targeting, 19*, 475–486. https://doi.org/10.3109/1061186x.2010.526227
40. Zhao, M.-X., & Zhu, B.-J. (2016). The research and applications of quantum dots as nano-carriers for targeted drug delivery and cancer therapy. *Nanoscale Research Letters, 11*, 207. https://doi.org/10.1186/s11671-016-1394-9
41. Abbasi, Y. F., Bera, H., Cun, D., & Yang, M. (2023). Recent advances in PH/enzyme-responsive polysaccharide-small-molecule drug conjugates as nanotherapeutics. *Carbohydrate Polymers, 312*, 120797. https://doi.org/10.1016/j.carbpol.2023.120797
42. Samia, A. C. S., Dayal, S., & Burda, C. (2006). Quantum dot-based energy transfer: Perspectives and potential for applications in photodynamic therapy. *Photochemistry and Photobiology, 82*, 617. https://doi.org/10.1562/2005-05-11-ir-525
43. Dubertret, B. (2002). In vivo imaging of quantum dots encapsulated in phospholipid micelles. *Science, 298*, 1759–1762. https://doi.org/10.1126/science.1077194
44. Cai, W., Shin, D.-W., Chen, K., Gheysens, O., Cao, Q., Wang, S. X., Gambhir, S. S., & Chen, X. (2006). Peptide-labelled near-infrared quantum dots for imaging tumor vasculature in living subjects. *Nano Letters, 6*, 669–676. https://doi.org/10.1021/nl052405t
45. Li, X., Vinothini, K., Ramesh, T., Rajan, M., & Ramu, A. (2020). Combined photodynamic-chemotherapy investigation of cancer cells using carbon quantum dot-based drug carrier system. *Drug Delivery, 27*, 791–804. https://doi.org/10.1080/10717544.2020.1765431
46. Zahed, Z., Hadi, R., Imanzadeh, G., Ahmadian, Z., Shafiei, S., Zadeh, A. Z., Karimi, H., Akbarzadeh, A. H., Abbaszadeh, M., Ghadimi, L. S., et al. (2024). Recent advances in fluorescence nanoparticles "quantum dots" as gene delivery system: A review. *International Journal of Biological Macromolecules, 254*, 127802–127802. https://doi.org/10.1016/j.ijbiomac.2023.127802
47. Mondal, S., Raut, J., & Sahoo, P. (2023). Gene silencing and gene delivery in therapeutics: Insights using quantum dots. *Frontiers in Bioscience-Landmark, 28*, 364–364. https://doi.org/10.31083/j.fbl2812364
48. Luo, G., Long, J., Zhang, B., Liu, C., Ji, S., Xu, J., Yu, X., & Ni, Q. (2012). Quantum dots in cancer therapy. *Expert Opinion on Drug Delivery, 9*, 47–58. https://doi.org/10.1517/17425247.2012.638624
49. Weldon, B. A., Griffith, W. C., Workman, T., Scoville, D. K., Kavanagh, T. J., & Faustman, E. M. (2018). In vitro to in vivo benchmark dose comparisons to inform risk assess-

ment of quantum dot nanomaterials. *Wiley Interdisciplinary Reviews: Nanomedicine and Nanobiotechnology, 10*. https://doi.org/10.1002/wnan.1507
50. Lin, X., & Chen, T. (2023). A review of in vivo toxicity of quantum dots in animal models. *International Journal of Nanomedicine, 18*, 8143–8168. https://doi.org/10.2147/ijn.s434842
51. Li, Y., Zhang, T., & Tang, M. (2021). Toxicity of quantum dots on target organs and immune system. *Journal of Applied Toxicology, 42*, 17–40. https://doi.org/10.1002/jat.4180
52. Eşref, D., & Vincent, C. (2017). Evaluation of the potential genotoxicity of quantum dots. A review. *Aspects of Nanotechnology, 1*. https://doi.org/10.36959/758/536
53. Schipper, M. L., Cheng, Z., Lee, S.-W., Bentolila, L. A., Iyer, G., Rao, J., Chen, X., Wu, A. M., Weiss, S., & Gambhir, S. S. (2007). Micro PET-based biodistribution of quantum dots in living mice. *Journal of Nuclear Medicine, 48*, 1511–1518. https://doi.org/10.2967/jnumed.107.040071
54. Benner, D., Yadav, P., & Bhatia, D. (2023). Red emitting carbon dots: Surface modifications and bioapplications. *Nanoscale Advances, 5*, 4337–4353. https://doi.org/10.1039/D3NA00469D
55. Soo Choi, H., Liu, W., Misra, P., Tanaka, E., Zimmer, J. P., Itty Ipe, B., Bawendi, M. G., & Frangioni, J. V. (2007). Renal clearance of quantum dots. *Nature Biotechnology, 25*, 1165–1170. https://doi.org/10.1038/nbt1340
56. Bourquin, J., Milosevic, A., Hauser, D., Lehner, R., Blank, F., Petri-Fink, A., & Rothen-Rutishauser, B. (2018). Biodistribution, clearance, and long-term fate of clinically relevant nanomaterials. *Advanced Materials, 30*, 1704307. https://doi.org/10.1002/adma.201704307
57. Mahendra, S., Zhu, H., Colvin, V. L., & Alvarez, P. J. (2008). Quantum dot weathering results in microbial toxicity. *Environmental Science & Technology, 42*, 9424–9430. https://doi.org/10.1021/es8023385
58. Tufani, A., Qureshi, A., & Niazi, J. H. (2021). Iron oxide nanoparticles based Magnetic Luminescent Quantum Dots (MQDs) synthesis and biomedical/biological applications: A review. *Materials Science and Engineering: C, 118*, 111545. https://doi.org/10.1016/j.msec.2020.111545
59. Rajappan-Achary, S., Agouram, S., Reig, C., Sánchez-Royo, J. F., Martínez-Tomás, M. C., & Muñoz-Sanjosé, V. (2011). Self-assembled zinc oxide quantum dots using spray pyrolysis methodology. *Crystal Growth & Design, 11*, 3790–3801. https://doi.org/10.1021/cg2003113
60. Hardman, R. (2006). A toxicologic review of quantum dots: Toxicity depends on physicochemical and environmental factors. *Environmental Health Perspectives, 114*, 165–172. https://doi.org/10.1289/ehp.8284
61. Choi, H. S., Liu, W., Liu, F., Nasr, K., Misra, P., Bawendi, M. G., & Frangioni, J. V. (2009). Design considerations for tumour-targeted nanoparticles. *Nature Nanotechnology, 5*, 42–47. https://doi.org/10.1038/nnano.2009.314
62. Derfus, A. M., Chan, W. C. W., & Bhatia, S. N. (2004). Probing the cytotoxicity of semiconductor quantum dots. *Nano Letters, 4*, 11–18. https://doi.org/10.1021/nl0347334
63. Gidwani, B., Sahu, V., Shukla, S. S., Pandey, R., Joshi, V., Jain, V. K., & Vyas, A. (2021). Quantum dots: Prospectives, toxicity, advances and applications. *Journal of Drug Delivery Science and Technology, 61*, 102308. https://doi.org/10.1016/j.jddst.2020.102308
64. Etefa, H. F., Tessema, A. A., & Dejene, F. B. (2024). Carbon dots for future prospects: Synthesis, characterizations and recent applications: A review (2019–2023). *C, 10*, 60. https://doi.org/10.3390/c10030060
65. Barhoum, A., Meftahi, A., Sadat, M., Ehsan, M., & Alem, F. (2023). A review on carbon dots as innovative materials for advancing biomedical applications: Synthesis, opportunities, and challenges. *Journal of Materials Science, 58*, 13531–13579. https://doi.org/10.1007/s10853-023-08797-6
66. Balalaeva, I. V., Zdobnova, T. A., Krutova, I. V., Brilkina, A. A., Lebedenko, E. N., & Deyev, S. M. (2012). Passive and active targeting of quantum dots for whole-body fluorescence imaging of breast cancer xenografts. *Journal of Biophotonics, 5*, 860–867. https://doi.org/10.1002/jbio.201200080

67. Gaur, M., Misra, C., Yadav, A. B., Swaroop, S., Ó Maolmhuaidh, F., Bechelany, M., & Barhoum, A. (2021). Biomedical applications of carbon nanomaterials: Fullerenes, quantum dots, nanotubes, nanofibers, and graphene. *Materials, 14*, 5978. https://doi.org/10.3390/ma14205978
68. Zhou, W., Hu, Z., Wei, J., Dai, H., Chen, Y., Liu, S., Duan, Z., Xie, F., Zhang, W., & Guo, R. (2022). Quantum dots-hydrogel composites for biomedical applications. *Chinese Chemical Letters, 33*, 1245–1253. https://doi.org/10.1016/j.cclet.2021.09.027
69. Filali, S., Pirot, F., & Miossec, P. (2020). Biological applications and toxicity minimization of semiconductor quantum dots. *Trends in Biotechnology, 38*, 163–177. https://doi.org/10.1016/j.tibtech.2019.07.013
70. Jamieson, T., Bakhshi, R., Petrova, D., Pocock, R., Imani, M., & Seifalian, A. M. (2007). Biological applications of quantum dots. *Biomaterials, 28*, 4717–4732. https://doi.org/10.1016/j.biomaterials.2007.07.014
71. Abdellatif, A. A., Younis, M. A., Alsharidah, M., Al Rugaie, O., & Tawfeek, H. M. (2022). Biomedical applications of quantum dots: Overview, challenges, and clinical potential. *International Journal of Nanomedicine, 17*, 1951–1970. https://doi.org/10.2147/ijn.s357980
72. Sahoo, S.; Nayak, A.; Gadnayak, A.; Sahoo, M.; Dave, S.; Mohanty, P.; Mohanty, J.N.; Das, J Quantum dots enabled point-of-care diagnostics: A new dimension to the nanodiagnosis. Available online: https://www.sciencedirect.com/science/article/abs/pii/B9780323857253000052. Accessed on 31 July 2024.

Biomedical Applications and Toxicity Mitigation of Functional Quantum Dots

Bhoomika Sridhar, Merla Sudha, Suvarna Ghosh, Belaguppa Manjunath Ashwin Desai, and Pronama Biswas

1 Introduction

Quantum dots (QDs), discovered in 1981, are nanoscale semiconductor particles (less than 10 nm) with unique optical properties, such as size-tunable fluorescence, high brightness, and exceptional photostability, making them highly suitable for a variety of applications. Owing to their excellent electrochemical properties, these features make QDs ideal for applications in optoelectronic devices, such as QLEDs and solar cells, as well as in sensors and batteries. Their small size allows them to penetrate biological membranes, making them valuable for cellular imaging and targeted drug delivery. QDs enable high-resolution imaging and multianalyte detection in biomedical applications, including cancer detection and infection monitoring. However, challenges such as potential toxicity and environmental impact must be addressed. Researchers are working on improving the biocompatibility of QDs and exploring safer alternatives to maximize their benefits while minimizing adverse effects. QDs offer significant advantages over traditional contrast agents. Their size-tunable fluorescence supports multiplexed imaging, enabling the simultaneous high-resolution visualization of multiple biological targets. This capability addresses the limitations of organic dyes and fluorescent proteins, which are often plagued by issues such as photobleaching and narrow spectral ranges. By emitting light across a broader spectrum, QDs enhance their imaging capabilities, providing a more detailed and comprehensive view of biological processes. They play a crucial role by improving therapeutic precision through both passive and active targeting

B. Sridhar · M. Sudha · S. Ghosh · P. Biswas (✉)
School of Basic and Applied Sciences, Dayananda Sagar University, Bengaluru, India
e-mail: pronama-sbas@dsu.edu.in

B. M. A. Desai
School of Engineering, Dayananda Sagar University, Bengaluru, India

S. P. Kabekkodu et al. (eds.), *Functional Quantum Dots*, Lecture Notes in Nanoscale Science and Technology 40,
https://doi.org/10.1007/978-3-031-98219-4_4

mechanisms. Passive targeting leverages the enhanced permeability and retention (EPR) effect, which allows nanoparticles to accumulate in tumor tissues due to their leaky vasculature. This effect can be further optimized through various delivery–release systems, including pH-sensitive, photothermal, and magnetic thermal modes. Active targeting involves functionalizing QDs with specific ligands that bind to cell receptors, ensuring targeted delivery of therapeutic agents while minimizing off-target effects. These approaches highlight the versatility of QDs in advancing targeted drug delivery systems. QDs also contribute significantly to biosensing technologies, offering high sensitivity and specificity in detecting and analysing biological molecules [1]. QD-based biosensors include photoluminescent, electrochemical, and electrochemiluminescence types. Photoluminescent biosensors utilize QDs' stable fluorescence for detecting interactions such as protein or nucleic acid presence. Electrochemical biosensors, which are enhanced by the electron-mediating properties of QDs, improve the sensitivity of analytes such as glucose or DNA. Electrochemiluminescence biosensors combine electrochemical and luminescent detection to provide highly sensitive and specific analyses [2].

Despite their advantages, the integration of QDs into clinical and diagnostic settings raises concerns about their potential toxicity and environmental impact. QDs can induce toxicity at the cellular and organ levels through mechanisms such as the release of toxic cadmium ions, oxidative stress, and mitochondrial apoptosis. These effects can lead to inflammation and damage in organs such as the liver and kidneys, as well as disruption of the blood–brain barrier and neuronal tissue. To address these challenges, researchers are exploring various strategies to increase the biocompatibility and reduce the toxicity of QDs [1]. Approaches include encapsulation to stabilize QDs, surface coating to minimize adverse effects, size optimization to control biodistribution, and dosage reduction to maintain efficiency while reducing the concentration and surface charge manipulation to improve interactions with biological systems. By addressing these issues, researchers aim to maximize the benefits of QDs while ensuring their safe and effective use in biomedical applications [3].

This chapter introduces QDs, nanoscale semiconductor particles known for their unique optical properties, making them ideal for applications in optoelectronics, sensors, and biomedical fields, highlighting how QDs enhance imaging, drug delivery, and biosensing due to their size-tunable fluorescence and high photostability. The chapter also discusses potential toxicity and environmental impact challenges. In summary, this chapter presents ongoing research efforts to improve the biocompatibility of QDs, maximizing their benefits while minimizing adverse effects.

2 Biomedical Applications of Quantum Dots

QDs have emerged as versatile and powerful tools in the biomedical field, offering unique optical and electronic properties that enable a wide range of applications. These nanoscale semiconductor particles exhibit unique properties with narrow, symmetric emission bands. These characteristics make QDs particularly suitable for applications in bioimaging, biosensing, and drug delivery, where precise detection, tracking, and manipulation at the molecular and cellular levels are crucial. The integration of QDs into biomedical technologies has the potential to revolutionize diagnostics and therapeutics, providing enhanced sensitivity, specificity, and functionality compared with traditional methods.

In bioimaging, QDs are used as fluorescent probes for visualizing biological structures and processes with exceptional clarity and resolution. Their brightness and resistance to photobleaching allow for prolonged imaging sessions, making them ideal for tracking dynamic events in living cells and tissues. QDs can be conjugated with biomolecules such as antibodies, peptides, or nucleic acids, enabling targeted imaging of specific cellular components, disease markers, or genetic sequences. This targeted imaging capability facilitates early diagnosis and monitoring of diseases, such as cancer, by allowing clinicians to observe the molecular changes associated with disease progression in real time [1].

In addition to imaging, QDs play crucial roles in biosensing and drug delivery. In biosensing, QDs are employed to develop highly sensitive and specific assays for detecting biomolecules, pathogens, and environmental toxins. Their unique optical properties enable the creation of sensors that can provide rapid, accurate, and multiplexed detection, which is essential for early diagnosis and disease management. In drug delivery, QDs are engineered to carry therapeutic agents and deliver them to specific target sites within the body. Their ability to be functionalized with targeting ligands and their excellent imaging capabilities allow for precise delivery and real-time tracking of therapeutic interventions. As a result, QDs hold significant promise for advancing personalized medicine, enabling the development of tailored treatment strategies that maximize therapeutic efficacy while minimizing side effects [4].

2.1 *Quantum Dots for Advanced Bioimaging*

Bioimaging involves the visualization of biological processes in real time, allowing for detailed observation of cellular and molecular activities. The superior properties of QDs, including their broad excitation spectra and narrow, size-tunable emission spectra, make them ideal candidates for bioimaging applications. QDs have significantly advanced both fluorescent and multimodal imaging techniques. Fluorescent imaging harnesses the bright, stable emission of QDs across a spectrum of wavelengths, allowing for the visualization of biological processes with high resolution and depth. This capability is particularly beneficial in the precise imaging and diagnosis of tumors, where QDs conjugated with antibodies can specifically bind to cancer cells. Furthermore, QDs enable multimodal imaging by incorporating properties suitable for magnetic resonance imaging (MRI), computed tomography (CT), and positron emission tomography (PET), alongside fluorescence. This multimodal approach provides comprehensive diagnostic information, facilitating more accurate and detailed imaging of disease states and improving therapeutic outcomes. The versatility of QDs in bioimaging underscores their potential to transform the landscape of medical diagnostics, enabling early detection and precise monitoring of diseases.

2.1.1 Fluorescence Imaging Using Quantum Dots

Fluorescent imaging is a powerful technique that uses fluorescent probes to label and visualize biological molecules or structures. QDs offer significant advantages over traditional fluorescent dyes and proteins, including superior photostability, brightness, and the ability to tune their emission wavelengths by changing their size. Owing to their high quantum yield, QDs are exceptionally bright, meaning that they can absorb and emit light very efficiently. This brightness, coupled with their resistance to photobleaching, allows for long-term imaging of biological samples without significant loss of signal. This is particularly useful in time-lapse imaging, where prolonged observation of dynamic processes is necessary. The emission wavelength of QDs can be precisely controlled by varying their size. Smaller QDs emit light at shorter wavelengths (towards the blue end of the spectrum), whereas larger QDs emit light at longer wavelengths (towards the red end of the spectrum). This tunability allows for multiplexed imaging, where multiple QDs with different emission colors can be used simultaneously to label different targets within the same sample. QDs have broad absorption spectra, which means that they can be excited by a wide range of wavelengths. However, they emit light in very narrow, specific bands. This property reduces the spectral overlap between different QD probes, enhancing the clarity and contrast of multicolor imaging [1].

They can be conjugated to antibodies, peptides, or other targeting molecules to specifically bind to and label cells, tissues, or subcellular structures. This targeted labelling allows for precise visualization of the location and distribution of specific biomolecules in biological samples. The exceptional photostability of QDs makes them ideal for single-molecule imaging, where the goal is to observe the behavior of individual biomolecules over time. This can provide valuable information about molecular interactions and dynamics at the nanoscale.

One of the applications of QDs in fluorescence imaging is their use in tracking and visualizing cancer cells. For example, a study conducted by Gao et al., 2004 demonstrated the use of QDs to target and image prostate cancer cells in live mice. In this study, QDs were conjugated to antibodies specific to prostate-specific membrane antigen (PSMA), a biomarker that is overexpressed in prostate cancer cells. When injected into mice with prostate tumors, these QD–antibody conjugates are specifically bound to cancer cells, allowing researchers to visualize and track tumor growth and spread via fluorescence imaging. The exceptional brightness and stability of the QDs provided clear, high-resolution images over extended periods, highlighting their potential for long-term tracking of cancer progression in vivo [5]. Michalet et al., 2009 used QDs for single-molecule imaging to observe the dynamics of membrane receptors on living cells. Researchers have attached QDs to specific proteins on the cell surface, enabling real-time tracking of individual protein molecules as they move and interact within the cell membrane. The high photostability of the QDs was crucial in this application, as it allowed continuous observation of the proteins without photobleaching, providing insights into the mechanisms of receptor-mediated signalling processes at the molecular level. This type of single-molecule tracking is invaluable for understanding the intricate behaviors of proteins and their roles in cellular functions [6]. Similarly, near-infrared QDs (NIR QDs) and a membrane-impermeable etchant were used to achieve highly tumor-specific signals via in vivo imaging. ZHS-QDs, which are composed of zinc, mercury, selenium, and sulfur, were coated with PEG to reduce mononuclear phagocyte system uptake. In mice with orthotopic breast and pancreatic tumors, QDs are delivered via the iRGD peptide, followed by selective etching to quench excess QDs, resulting in a tumor-specific signal without notable toxicity. Additionally, in peritoneal tumors, iRGD or PBS with ZHS-QDs was coinjected, demonstrating high specificity in detecting peritoneal tumors (Fig. 1). This method holds promise for enhancing tumor imaging specificity [7].

2.1.2 Multimodal Imaging Capabilities

Multimodal imaging combines two or more imaging techniques to provide complementary information about a biological sample. QDs are particularly well suited for multimodal imaging because of their ability to be engineered with various functional properties beyond fluorescence, including magnetic and photoacoustic properties. QDs can be synthesized to incorporate multiple functionalities, such as combining fluorescence with magnetic properties. This allows them to be used in both optical imaging and MRI, providing a more comprehensive view of the biological specimen. By integrating different imaging modalities, QDs can help overcome the limitations of individual techniques. For example, combining fluorescence

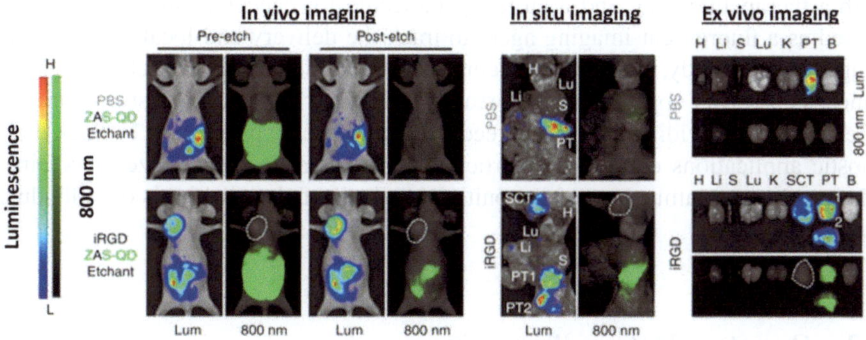

Fig. 1 Mice bearing peritoneal tumors (PTs) induced with MKN45P-luc luciferase-positive human gastric cancer cells were administered an i.p. coinjection of iRGD or PBS with ZHS-QDs. Live in vivo and postnecropsy in situ images of the tumor-bearing mice, along with ex vivo images of resected tissues, are displayed. Lum, luminescence; 800 nm, NIR (near-infrared region). The white dotted line indicates a subcutaneous tumor (SCT). NIR signal per unit area in collected tissues. Reprinted with permission from [7]

imaging with photoacoustic imaging can improve both the spatial resolution and depth of imaging, allowing for clearer and more detailed visualization of deep tissues. Multimodal QDs enable real-time monitoring of biological processes via multiple imaging techniques simultaneously. This approach can provide more dynamic and complete information about physiological and pathological events, such as tracking the delivery and release of drugs at the target site in real time [1]. A notable example is the development of QD-based probes for combined fluorescence MRI. In a study by Santra et al., 2005, QDs were engineered with magnetic cores to create fluorescent magnetic nanoparticles. These hybrid nanoparticles could be used for both fluorescence imaging to visualize cells and tissues and MRI to obtain detailed anatomical information. When used in animal models, these probes allow researchers to track the distribution and accumulation of the nanoparticles in the body via MRI while simultaneously visualizing their specific interactions with target cells through fluorescence. This dual-modality approach provided a more complete picture of the biodistribution and targeting efficiency of the nanoparticles, which is critical for their application in drug delivery and therapy [9].

QDs have also been combined with photoacoustic imaging to enhance the visualization of deep tissues (Fig. 2). In a few studies, QDs were integrated with gold nanoparticles to create probes that could be used for both fluorescence and photoacoustic imaging. The gold component of the probes enhanced their photoacoustic signal, allowing for deeper tissue penetration and improved imaging resolution. When injected into tumor-bearing mice, these probes provide detailed images of the tumors at depths that are challenging for conventional fluorescence imaging alone. The combination of fluorescence and photoacoustic imaging has enabled researchers to obtain high-resolution images of the tumor structure and vasculature, providing valuable insights into tumor biology and progression [10]. In the realm of theranostics, where diagnostic and therapeutic functions are combined, QDs have shown great promise. For example, Smith et al., 2008 utilized QDs conjugated with both a therapeutic drug and a targeting molecule for cancer treatment. The QDs served as a fluorescent imaging agent to track the delivery and localization of the drug within the body, while the drug component provided therapeutic effects against cancer cells. This approach allows real-time monitoring of drug distribution and effectiveness, providing immediate feedback on treatment progress [11]. Such theranostic applications of QDs are particularly beneficial in personalized medicine, where tailored treatments can be monitored and adjusted on the basis of individual responses.

2.2 Quantum Dots in Biosensing

Biosensing is the process of using biosensors for detecting and measuring biological analytes for theranostic purposes. It involves the use of different techniques that play crucial roles in healthcare by enabling rapid, sensitive, and specific detection of disease-related biomarkers. Multiple studies have shown the importance of

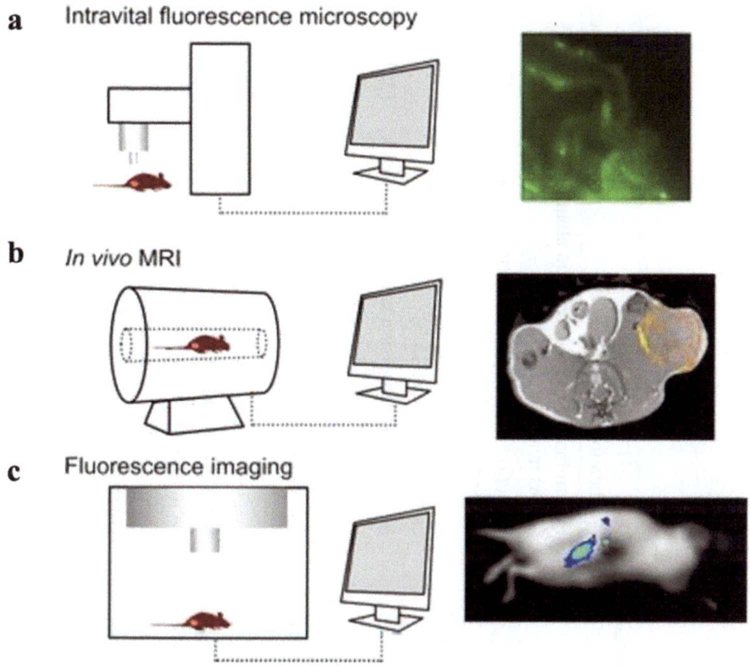

Fig. 2 (**a**) Multimodal imaging via quantum dots (QDs) via intravital fluorescence microscopy (IVM) visualizes QDs within superficial blood vessels, showing microvessels in tumor-bearing mice after intravenous injection. (**b**) Magnetic resonance imaging (MRI) provides high-resolution, three-dimensional localization of QDs in a mouse with a subcutaneous tumor. (**c**) Whole-body fluorescence imaging enables the detection of QDs throughout the mouse body, as illustrated by the fluorescence image of a mouse with a kidney tumor following intravenous administration. Reprinted with permission from [8]

biosensing in biomedical applications across different fields, focusing on the diverse range of materials and approaches used in biosensing [1]. QD biosensors utilize the optical and electronic characteristics of QDs, categorized on the basis of their mechanisms, detection modes, and applications. Photoluminescent biosensors rely on photoluminescence (PL), where QDs absorb photons and emit light [12]. Electrochemical (EC) biosensors are equipped with biomolecules such as enzymes or antibodies that bind specifically to target analytes to convert biochemical interactions into electrical signals [13]. ECL biosensors combine electrochemical and luminescent processes for sensitive analyte detection. By applying a potential to a modified electrode, reactive species are produced when an analyte binds, resulting in light emission [14]. These biosensors are vital for biomedical and clinical applications because of their sensitivity and optical versatility, ability to detect disease biomarkers such as cancer and facilitate treatment monitoring (Table 1). QD biosensors also enable real-time drug delivery tracking, precise cellular imaging for studying cell dynamics and protein locations, and applications in environmental

Table 1 Types of quantum dot (QD)-based biosensors and their applications

Types of biosensor	Nanomaterial	Conjugated/functionalized with	Applications	Reference
Photoluminescent biosensor	ZnS QDs	Mn^{2+}	Determination of bovine serum albumin (BSA) in aqueous solutions	[32]
	QODs	Ethylenediamine	Detection of *Staphylococcus aureus* and *Escherichiacoli*	[33]
Electrochemical biosensor	Thiolated graphene quantum dots (CysA/GQDs)	Gold nanospear (Au NSs)	Detection of breast cancer-specific carbohydrate (CA-15-3)	[34]
	GQDs	Polyaniline	Screening of depression markers-heat shock protein 70 (HSP70)	[35]
	CdTe/CdSe/ZnSe QDs	Themercaptopropionicacid(MPA)-	Detection of cholestrol	[26]
	GQDs	Multiwall carbon nanotubes (MWCNTs)	Detection of dopamine	[36]
	Nickel-cadmium quantum dots	Second antibody (Ab2) labelled probe	Determination of prostate-specific antigen (PSA) in human serum samples	[37]
Electrochemiluminescence biosensor	CdS QDs and GO-QDs	Chloramphenicol aptamer	Determination of chloramphenicol in samples	[31]
	CdSe QDs	DNA	Detection of carcinoembryonic antigen (CEA)	[38]
	Graphite phase carbon nitride QD (S-GCN QDs)	Sulfur	Quantification of the K-RAS gene	[39]

monitoring, DNA, and protein assays, highlighting their broad impact in the biomedical sciences [1].

2.2.1 Photoluminescent Biosensors

Photoluminescent biosensors are advanced tools that utilize light emission from specific materials to detect and quantify various analytes. Upon excitation by a light source, electrons in the QD are excited to a higher energy state. When these electrons return to their ground state, they produce PLs that emit light at a specific wavelength (Fig. 3a) [15]. The PL exhibited by QDs is size dependent, where the emission wavelength can be precisely controlled by adjusting the particle size. Larger QDs have smaller band gaps, resulting in the emission of light at longer wavelengths, whereas smaller QDs have larger band gaps and emit light at shorter wavelengths [1].

QDs are highly effective photoluminescent biosensors because of their core–shell structure, which enables them to emit light at specific wavelengths when excited by a light source. For this application, QDs are made as core–shell structures, usually cadmium selenide (CdSe) or cadmium telluride (CdTe) cores, with shells composed of zinc sulfide (ZnS). The shell enhances the quantum yield and photostability by passivating the surface of the core, reducing nonradiative recombination pathways [1]. Passivation involves coating the QD core surface to neutralize surface defects that can trap charge carriers caused by unsatisfactory chemical bonds. In the presence of surface defects, the excited electrons lose their energy to non-light-emitting processes, reducing the quantum yield [16]. Grandhi et al. demonstrated the effect of surface passivation on the PL efficiency of CdSe QDs, showing that modifying surface passivation with trioctylphosphine oxide (TOPO), trioctylphosphine (TOP), and alkylamine ligands can increase the PL efficiency from 15% to 40% [12]. The surface of QDs can also be functionalized with ligands, such as carboxyl, amine, or thiol groups, to facilitate the conjugation of biomolecules such as peptides or nucleic acids, allowing them to specifically bind target analytes [17] This binding event alters the photoluminescent properties of the QDs, such as changes in fluorescence intensity, wavelength shift, or fluorescence resonance energy transfer (FRET). These changes are measured via optical detection systems to identify and quantify the target analytes [18]. The high sensitivity, multiplexing capability, and long-term stability of QDs make them ideal for diverse applications in biological imaging and diagnostics. Multiplexing capacity is the ability to detect multiple analytes simultaneously with a single excitation wavelength [19].

FRET is a distance-dependent phenomenon where energy is transferred in a nonradiative manner from an excited-state donor fluorophore to an acceptor fluorophore through a dipole–dipole interaction (Fig. 4). This phenomenon depends on the spectral overlap between the emission spectra of the donor and acceptor [20]. In QD photoluminescent biosensors, QDs are utilized as energy donor and acceptor fluorophores, where FRET is applied to study molecular interactions by measuring

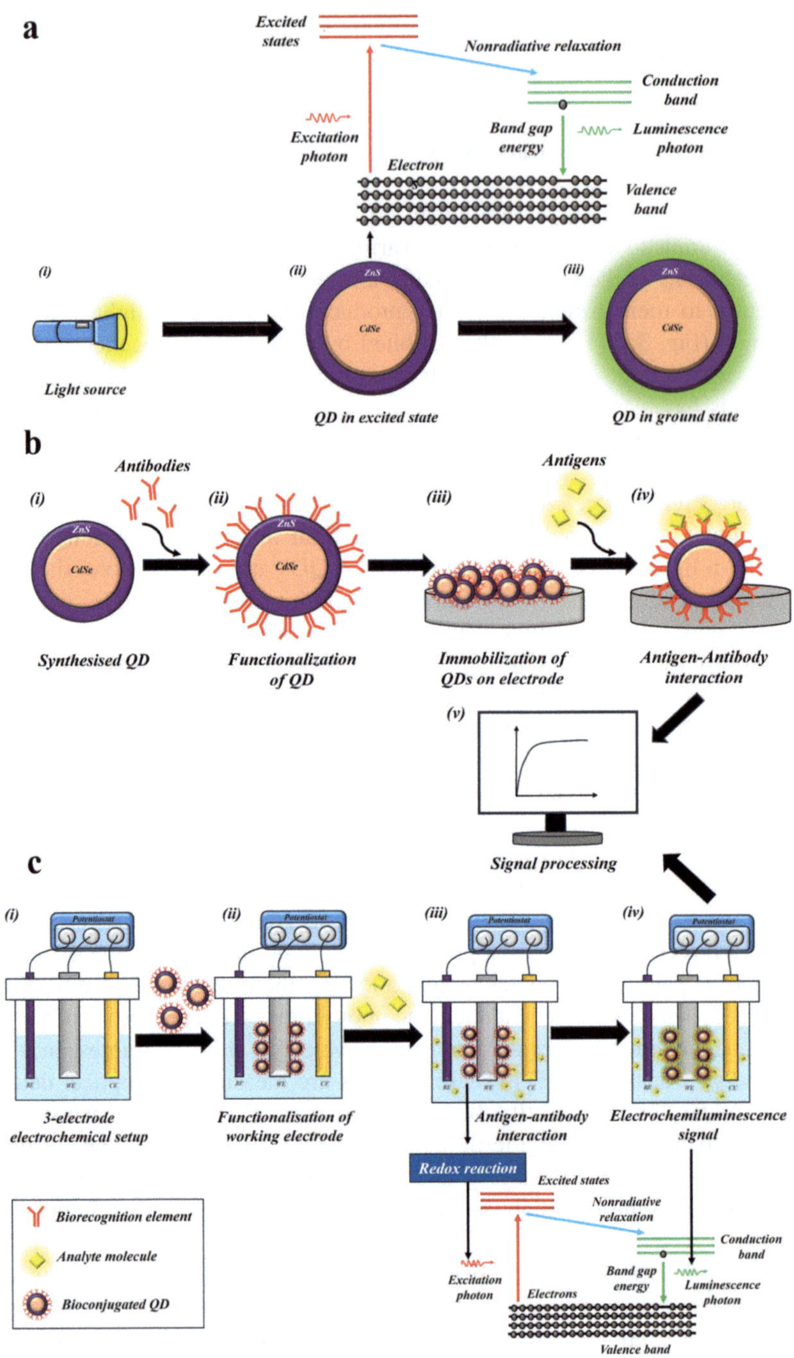

Fig. 3 Schematic representation illustrating the working principles of biosensors. (**a**) Photoluminescent biosensors are analytical devices that employ the principle of fluorescence emission to detect biomolecules present in the analyte. Upon absorption from a light source, the elec-

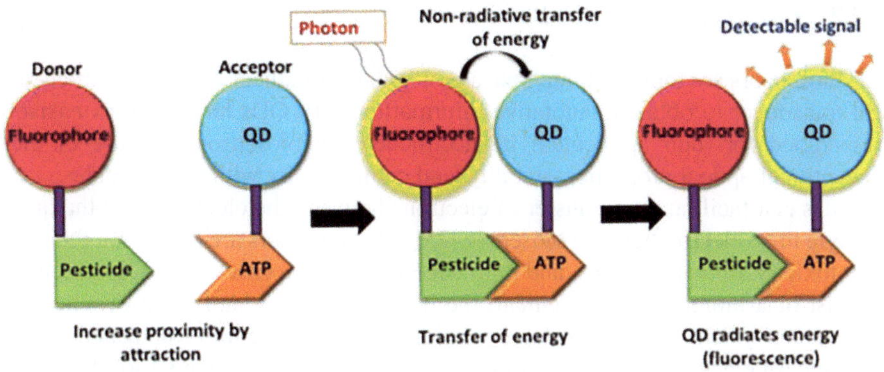

Fig. 4 Diagrammatic representation of fluorescence resonance energy transfer (FRET) imaging. A fluorophore transfers energy to quantum dots (QDs), resulting in measurable fluorescence. This technique allows for the detection and visualization of target compounds, such as ATP (adenosine triphosphate), via the fluorescence emitted from QDs. Reprinted with permission from [1]

changes in the QD fluorescence intensity, enabling sensitive and selective detection in biosensing applications. In a specific application, Gu, 2017 developed a highly sensitive, selective, and rapid method for detecting cell apoptosis, which combines the site-specific recognition and cleavage of the DEVD peptide with QD-based FRET [21].

The QDs in these biosensors are conjugated with biomolecules to increase their specificity for the target molecules in the analyte. For example, Lu et al., 2022 developed a biosensor with CdSe/CdS/ZnS QDs functionalized with aptamers for simultaneous detection of amyloid-β oligomers (AβO) and tau protein via FRET [22]. Aptamers are short strands of DNA or RNA oligonucleotides that can bind to target molecules with high affinity and selectivity [23].

Fig. 3 (Continued) trons present in quantum dots (QDs) are excited to a higher state and return to their ground state, resulting in photoluminescence (PL). (**b**) Electrochemical (EC) biosensors utilize QDs immobilized on electrode surfaces as transducers to convert biochemical interactions into electrical signals. When the target analyte is introduced, it binds to the functionalized QDs. This interaction induces alterations in the electrochemical characteristics of the QDs, including changes in current, potential, and impedance. (**c**) An electrochemiluminescence (ECL) biosensor consists of a three-electrode setup that includes a working electrode (WE), reference electrode (RE), and counter electrode (CE). The working electrode is coated with QDs on its surface, which are coupled with biomolecules to target specific analytes. The application of a potential between the working and reference electrodes initiates a redox reaction, causing the QDs to become excited. The excited QDs emit photons as they return to their ground state. The diagram was generated via Microsoft® PowerPoint® Version 16.86 (24060916) Build 16.0.16924.20054 (Microsoft®, Redmond, WA, USA)

2.2.2 Electrochemical (EC) Biosensors

EC biosensors are tools that integrate a biological component with an electrochemical transducer to obtain quantitative information [24]. QDs have been extensively researched for their application in electrochemical biosensors because of their exceptional optical properties, stability, and compatibility with biomolecules.

QDs can facilitate the transfer of electrons between the electrode and the target analyte in an electrochemical biosensor [25]. EC biosensors are known for their low cytotoxicity, optoelectronics, and high conductivity properties [13]. EC biosensors consist of a biorecognition element, the biosensor's core, and a physicochemical transducer system, which translates the biological response into a detectable signal for modern electrical instrumentation. Together, these components form the working electrode. Additionally, a reference electrode and a counter electrode are employed in electrical measurements [25]. In EC biosensors, QDs are immobilized on the surface of an electrode, which acts as the transducer and converts the biochemical interaction into an electrical signal. The target analyte binds to the functionalised QD when introduced into the system (Fig. 3b). These interactions lead to changes in the electrochemical properties of QDs, such as current, potential and impedance [26]. In recent years, the development of EC biosensors has undergone significant advancements through the incorporation of QDs, increasing their sensitivity and specificity for various analytes. One notable example is the work of Mokwebo et al., 2018, who developed an enzyme-based EC biosensor for cholesterol detection. This biosensor was synthesized by immersing a glassy carbon electrode/poly(propylene imine) dendrimer (GCE/PPI) electrode into a mercaptopropionic acid (MPA)-capped CdTe/CdSe/ZnSe QD solution, followed by incubation to allow the attachment of the amine group of the dendrimer [26]. The modified electrode was then incubated with cholesterol oxidase (ChOx), resulting in a biosensor capable of detecting cholesterol in the range of 0.1–10 mM with a sensitivity of 111.16 $\mu A\ mM^{-1}\ cm^{-2}$ and a detection limit (LOD) of 0.075 mM. Similarly, Cai et al., 2018 demonstrated the importance of functionalization in enhancing biosensor performance by synthesizing polyaniline-functionalized graphene QDs (PAGDs) to produce an EC sensor for the analysis of calycosin [27]. The functionalization with polyaniline not only improved the electrical conductivity but also provided specific binding sites, thereby increasing the sensitivity and selectivity of the biosensor. To further expand the application of functionalized nanomaterials, Arumugasamy et al., 2020 developed an EC biosensor for dopamine detection by incorporating graphene QDs (GQDs) with acid-functionalized multiwall carbon nanotubes (MWCNTs) on the surface of a glassy carbon electrode (GCE). This innovative combination improved the sensor performance, demonstrating the synergistic effects of nanomaterial hybrids in biosensor technology [28].

EC biosensors encompass a range of advanced technologies, including enhanced chemiluminescence (ECL) biosensors, which integrate electrochemical and luminescent detection methods to increase sensitivity and specificity in analytical applications. ECL is the phenomenon of the generation of light through electrochemical reactions [29]. An ECL biosensor consists of a working, reference and counterelectrode [30]. ECL reaction at the working electrode functionalized with QDs on the surface conjugated with biomolecules for targeting specific analytes. Redox occurs when a potential is applied across the working and reference electrodes, leading to the generation of QDs with excited states. These excited QDs emit photons when they revert to their ground state (Fig. 3c) [14]. The intensity of the electrochemiluminescence signal is directly proportional to the concentration of the target molecules present in the analyte [14].

In ECL biosensing, QDs are often used to increase the sensitivity and specificity of detection by leveraging their unique photophysical properties. These QDs can be functionalized and combined with various biomolecules to create sensitive biosensors for detecting a range of analytes. For example, Jung et al., 2016 demonstrated the application of ECL biosensors by synthesizing a biosensor functionalized with TGA-capped CdSe QDs and immobilized with glucose oxidase (GOD). This setup enabled the detection of low-level glucose, revealing how QDs can be used in ECL biosensing to improve glucose detection capabilities [30]. Similarly, He et al., 2020 developed a biosensor by synthesizing CdS QDs and attaching them to graphene oxide (GO) with poly(diallyldimethylammonium chloride) (PDDA). They characterized these GO-QDs via UV–visible absorption spectroscopy, fluorescence spectroscopy, and transmission electron microscopy. The GO-QDs were then applied to a glassy carbon electrode (GCE). To detect chloramphenicol, they used a biotin-modified aptamer that binds specifically to chloramphenicol and pairs it with its complementary DNA strand (cDNA) to form double-stranded DNA (dsDNA) on the GO-QD-modified electrode. When chloramphenicol binds to the aptamer, it causes the dsDNA to unwind and detach from the electrode. This detachment leads to an increased ECL signal because less hydrogen peroxide is consumed. This example shows how integrating QDs with GO and aptamers can increase the sensitivity and specificity of biosensors [31].

3 Quantum Dots as Theranostic Tools

Owing to the remarkable optical properties of QDs, they are prime candidates for bioimaging and diagnostic applications. By harnessing their expansive surface-to-volume ratio, sophisticated multifunctional nanoplatforms have been crafted, allowing QDs to serve as both imaging agents and nanoscaffolds for diverse therapeutic and theranostic uses. Recent studies have highlighted the potential of QD-based theranostic systems for cancer management. One investigation reported the use of biocompatible Fe_3O_4-ZnO core–shell magnetic QDs that could simultaneously visualize and treat tumors, leveraging their ability to generate singlet oxygen and exhibiting superparamagnetism for MRI-guided photodynamic therapy [40]. An alternative approach employs CdTe QD-based nanocapsules with a positively charged oily core loaded with anticancer drugs enveloped in a chondroitin sulfate coating to enhance tumor targeting [41]. Furthermore, an ingenious strategy using the iron-binding cationic protein lactoferrin was found to induce an on/off effect on QD fluorescence, initially dampening it and then restoring it upon bond cleavage within tumor cells, demonstrating the versatile theranostic capabilities of QDs beyond their conventional applications [42].

In a significant study, researchers developed multifunctional QD-based nanoprobes for the simultaneous detection and treatment of breast cancer. This study utilized QDs conjugated with targeting ligands such as antibodies or peptides that specifically bind to breast cancer cells. These QD conjugates were also loaded with chemotherapeutic drugs, providing a dual function of imaging and therapy. When these QD nanoprobes bind to cancer cells, they emit fluorescence, enabling real-time imaging of the tumor. Moreover, chemotherapeutic drug release can be triggered by external stimuli such as pH changes or light, leading to targeted destruction of cancer cells. This study demonstrated high specificity and sensitivity in targeting breast cancer cells, confirming the efficacy of QDs as potent theranostic agents in oncology [5]. Scientists have developed QDs conjugated with prostate cancer-specific peptides, enabling targeted imaging and therapy. The QDs were used to track the distribution and localization of the nanoparticles within the tumor tissues via their fluorescent properties. Additionally, the QDs were loaded with chemotherapy drugs, which were released in response to the acidic environment of the tumor. This targeted approach minimizes the side effects typically associated with chemotherapy and enhances the therapeutic efficacy against prostate cancer [43]. A similar study explored the use of QDs for imaging and treating brain tumors. Researchers have designed QDs coated with polyethylene glycol (PEG) to enhance their biocompatibility and stability in biological environments. These QDs were further conjugated with tumor-specific ligands that target brain tumor cells. In vivo experiments revealed that these QDs could cross the blood–brain barrier, which is a significant challenge in brain cancer treatment. Upon reaching the tumor site, the QDs enabled precise imaging due to their strong fluorescent properties. Additionally, when coupled with therapeutic agents, QDs facilitate targeted drug delivery, significantly reducing tumor size in animal models. This study highlighted the potential of QDs in overcoming critical challenges in brain cancer theranostics [44].

4 Advancing Drug Delivery with Quantum Dots

Following advancements in bioimaging, biosensing, and the integration of QDs as theranostic agents, targeted drug delivery has emerged as a pivotal innovation in the realm of precision medicine. Targeted drug delivery systems are designed to direct therapeutic agents specifically to diseased tissues, minimizing systemic exposure and reducing side effects. This approach leverages the unique molecular signatures of pathological sites, utilizing ligands, antibodies, or peptides that can bind precisely to disease markers. By enhancing the accumulation of drugs at the target site, these systems not only improve therapeutic efficacy but also mitigate adverse effects, thereby offering a significant leap forward in the treatment of complex diseases such as cancer, cardiovascular disorders, and infectious diseases. The development of targeted drug delivery systems represents an amalgation of advanced materials science, nanotechnology, and molecular biology, promising for transforming the therapeutic landscape with greater accuracy and effectiveness.

4.1 Passive Targeting

Passive targeting utilizes the inherent characteristics of the disease microenvironment, such as leaky vasculature and altered pH, to accumulate therapeutic agents at the target site without the need for specific interactions between the drug carrier and the target cells. This phenomenon is particularly pronounced in solid tumors, where the EPR effect allows nanoparticles to preferentially accumulate within the tumor aperture because of their abnormal vasculature. Owing to their unique optical properties and nanoscale size, QDs are increasingly used in passive targeting. Their fluorescence allows for precise tracking of the distribution and accumulation of therapeutic agents in real time, enabling detailed imaging of the drug delivery process [45].

Furthermore, QDs can be engineered to respond to specific stimuli within the disease microenvironment, such as pH changes or enzymatic activity, to achieve controlled drug release. This ensures that the therapeutic agents are released only at the target site, reducing systemic toxicity [46]. The ability of QDs to combine diagnostic and therapeutic functions into a single platform makes them ideal candidates for passive targeting in various diseases. The development of QDs for passive targeting not only enhances the efficacy of drug delivery but also provides invaluable insights into the pharmacokinetics and biodistribution of nanomedicines, paving the way for more effective and safer therapeutic strategies.

4.1.1 Enhanced Permeability Retention (EPR)—pH Mode

The EPR-pH mode is a fundamental concept in cancer therapy. Tumors are characterized by their abnormal vasculature, which results in increased permeability, and impaired lymphatic system, resulting in compromised drainage, leading to the accumulation of nanoparticles such as QDs. This phenomenon is exploited to deliver drugs more effectively to the tumor site (Fig. 5a). This mode is particularly advantageous because it harnesses the natural characteristics of tumor biology, ensuring that drug release occurs precisely where it is needed. QDs can be engineered to be stable at physiological pH but to release their therapeutic payload in the acidic microenvironment typical of tumors [46]. For example, ZnO QDs stabilized with polyethylene glycol (PEG) and targeted with hyaluronic acid can effectively deliver drugs to cancer cells. The acidic environment triggers the release of the drug, ensuring that it acts specifically within the tumor while minimizing systemic toxicity [46].

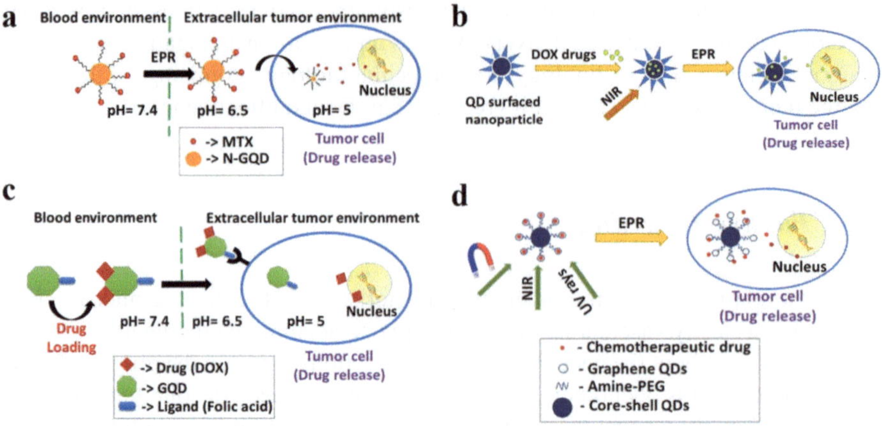

Fig. 5 Types of passive targeted drug delivery. (**a**) EPR-pH mode involves delivering drugs in response to the acidic environment near tumors, taking advantage of the effect. (**b**) EPR-photothermal delivery-release mode uses near-infrared to load the drug onto quantum dots (QDs). (**c**) The ligand-pH delivery-release mode employs a ligand on the QD that selectively binds to tumor cells to release the drug. (**d**) The photothermal/magnetic thermal delivery-release mode utilizes methods such as magnetic, photothermal, and radiation for drug loading. Reprinted with permission from [1]

Other studies have also shown that ZnO QDs loaded with doxorubicin have increased uptake and drug release under acidic conditions in the tumor microenvironment, leading to improved therapeutic outcomes. Moreover, the fluorescent properties of QDs allow real-time tracking and imaging, providing valuable insights into the distribution and release patterns of the drug [47]. However, the variability in the EPR effect among different tumor types presents a challenge. Not all tumors exhibit the same degree of permeability, which can affect the efficiency of drug delivery. Additionally, the potential for nonspecific accumulation in other areas with high vascular permeability, such as inflamed tissues, must be carefully managed. Despite these challenges, the EPR-pH mode remains a promising strategy for enhancing the precision and effectiveness of cancer therapies using QDs [7].

4.1.2 Photothermal/Magnetic Thermal Delivery—Release Mode

The photothermal/magnetic thermal delivery-release mode leverages external stimuli to trigger drug release. QDs can absorb light, particularly near-infrared (NIR) light, and convert it into heat, which can then be used to release the drug at the target site. This approach is known as photothermal therapy [48]. Gold QDs, for instance, are highly effective photothermal agents because of their strong absorption of NIR light. When these QDs are irradiated with NIR light, they generate heat that can induce drug release (Fig. 5d) [49].

Magnetic thermal delivery uses magnetic nanoparticles that generate heat upon exposure to an alternating magnetic field. This method allows for controlled drug release through hyperthermia. Fe_3O_4-ZnO core–shell QDs have been used for this purpose, combining the magnetic properties of iron oxide with the drug-carrying capacity of ZnO. These QDs can be guided to the tumor site via an external magnetic field, and once in place, an alternating magnetic field induces hyperthermia and drug release [50].

The dual-mode strategy is particularly useful for treating deep-seated tumors that are not easily accessible by conventional means. By combining imaging and therapy, QDs facilitate a theranostic approach, enabling physicians to monitor the treatment in real time and adjust it as needed [51]. Studies have demonstrated that QDs such as CdTe and Fe_3O_4-ZnO can effectively deliver drugs and enhance therapeutic outcomes through controlled photothermal and magnetic thermal release mechanisms. However, precise control of the irradiation or magnetic field is crucial to avoid damage to surrounding healthy tissue [52].

4.1.3 Ligand-pH Delivery-Release Mode

The ligand-pH delivery-release mode combines the specificity of ligand–receptor targeting with the controlled release capabilities of pH-sensitive systems. QDs can be functionalized with ligands that specifically bind to receptors overexpressed on target cells, such as cancer cells. Once QDs are internalized via receptor-mediated endocytosis, the acidic conditions of the endosomal and lysosomal compartments trigger the release of the drug. This dual-targeting strategy enhances both the specificity and control of drug release (Fig. 5c) [53].

For example, QDs functionalized with folic acid can target cancer cells that overexpress folate receptors. Once internalized, the acidic environment inside the cells causes the release of the drug payload of the QDs. This method ensures that the drug acts primarily within the target cells, reducing off-target effects and improving therapeutic efficacy [54]. A notable example is the use of folate-conjugated CdTe QDs loaded with doxorubicin, which have shown increased uptake in folate receptor-positive tumor cells and effective drug release in the acidic intracellular environment. These compounds are paired with fluorescent probes and can also be used in bioimaging [55]. This approach also addresses the challenge of drug resistance in cancer therapy. Ensuring that the drug is released only within the target cells minimizes the likelihood of drug efflux, a common mechanism of resistance. Additionally, the use of QDs allows for simultaneous imaging and therapy, providing a powerful tool for monitoring treatment progress. However, the stability of ligand–drug conjugates in circulation and the potential for immune recognition remain challenges that need to be addressed to fully realize the potential of this strategy [56].

4.1.4 EPR-Photothermal Model

The EPR-photothermal model combines the benefits of the EPR effect with those of photothermal therapy to enhance targeted drug delivery. QDs accumulate in tumors via the EPR effect and are then activated by light to generate heat, which facilitates drug release and induces hyperthermia to kill cancer cells. This hybrid approach takes advantage of the QDs' optical properties to achieve a synergistic effect, improving both the precision and effectiveness of cancer treatment (Fig. 5b).

In this model, QDs such as CdTe are encapsulated with chemotherapeutic agents such as rapamycin and celecoxib. These QDs accumulate in the tumor tissue due to the EPR effect. Upon exposure to NIR light, the QDs generate heat that not only enhances drug release but also directly damages tumor cells through hyperthermia. This combined effect can significantly improve therapeutic outcomes, as demonstrated in studies where QDs have been used to target and treat tumors effectively [46]. The EPR-photothermal model also allows real-time monitoring of the treatment process. The fluorescent properties of QDs enable imaging of the tumor and the distribution of the drug, providing valuable feedback that can be used to adjust the treatment in real time. This approach minimizes systemic toxicity and enhances the therapeutic index of chemotherapeutic agents. However, precise control over the irradiation process is essential to avoid damage to surrounding healthy tissues, and the variability of the EPR effect among different tumors must be considered [57, 58].

4.2 Active Targeting

Active targeting represents a sophisticated approach to drug delivery, where drug carriers are modified with molecules that specifically interact with target cells, significantly increasing the selectivity and efficacy of therapeutic agents. This strategy ensures that drugs are delivered precisely to diseased cells while minimizing exposure to healthy tissues. QDs play a pivotal role in active targeting because of their unique ability to be conjugated with a variety of targeting ligands, such as antibodies, peptides, and small molecules. Their excellent imaging capabilities further enable real-time tracking and monitoring of the drug delivery process, ensuring accurate localization and release of therapeutic agents [5].

Within the realm of active targeting, several innovative strategies have emerged, including exosome-mediated, folate-mediated, RNAi molecule-mediated, and glucose-mediated delivery. Exosome-mediated delivery leverages the natural ability of exosomes to transfer therapeutic molecules between cells, with QDs enhancing imaging and tracking [59]. Folate-mediated delivery utilizes folic acid to target cancer cells with overexpressed folate receptors, enabling targeted drug accumulation and imaging [60]. RNAi molecule-mediated delivery employs RNA interference for gene silencing, with QDs facilitating precise delivery and monitoring gene silencing efficiency [61]. Finally, glucose-mediated delivery exploits the high glucose uptake of certain cancer cells, using QDs linked with glucose to monitor and ensure targeted therapeutic delivery [62]. These approaches collectively harness the versatility of QDs, providing a comprehensive solution for advanced and precise drug delivery systems.

Active targeting offers several advantages over passive targeting, primarily through its enhanced specificity and precision. By modifying drug carriers with targeting ligands, active targeting can achieve a higher concentration of therapeutic agents at the disease site, reducing off-target effects and systemic toxicity. This is particularly beneficial in treating heterogeneous diseases such as cancer, where precise delivery to malignant cells is crucial. In contrast, passive targeting relies on the inherent characteristics of the disease microenvironment, such as leaky vasculature and altered pH, to accumulate therapeutic agents. While this approach can be effective, it often lacks precision in active targeting and may result in lower drug accumulation at the target site. However, the development of targeting ligands and their conjugation to drug carriers can be complex and costly in active targeting. The choice between active and passive targeting ultimately depends on the specific disease context, the availability of targetable markers, and the desired balance between precision and simplicity in drug delivery. The use of QDs in both strategies further enhances these approaches by enabling precise tracking and controlled drug release [4, 63].

4.2.1 Exosome-Mediated Targeted Delivery

Exosome-mediated targeted delivery utilizes exosomes loaded with QDs and therapeutic agents. These exosomes can be engineered to express specific ligands on their surface, enabling them to target particular cells or tissues. QDs within the exosomes allow real-time tracking and imaging, ensuring that the therapeutic agents are delivered accurately to the target site. This method is highly biocompatible and has low immunogenicity, making it a promising approach for targeted drug delivery [59]. For example, QDs can be loaded into exosomes derived from cells that naturally produce these vesicles. Exosomes can be engineered to express surface ligands that target specific cancer cells, ensuring the precise delivery of therapeutic agents. This method has been shown to improve the targeting efficiency and therapeutic outcomes of anticancer drugs.

Moreover, the use of QDs enables the tracking of exosomes in real time, providing valuable insights into their biodistribution and targeting efficacy [64]. Exosome-mediated delivery also offers the advantages of natural vesicle properties, such as stability in circulation and the ability to cross biological barriers. These properties increase delivery efficiency and reduce the potential for immune recognition. However, the production and purification of exosomes on a large scale remain challenging, and further research is needed to optimize this delivery system for clinical use.

4.2.2 Folate-Mediated Delivery (FA)

Folate-mediated delivery (FA) takes advantage of the overexpression of folate receptors in many cancer cells. QDs conjugated with folic acid can bind to these receptors, facilitating the targeted delivery of drugs. This method enhances the specificity of drug delivery, as seen in studies where QDs linked with folic acid and chemotherapeutic drugs such as doxorubicin have shown increased uptake in folate receptor-positive tumors, minimizing effects on normal cells.

For example, folate-conjugated CdTe QDs have been used to deliver doxorubicin to folate receptor-positive cancer cells. The QDs bind to the folate receptors on the cancer cells, leading to receptor-mediated endocytosis and internalization of the QDs. Once inside the cells, the acidic environment triggers the release of doxorubicin, ensuring that the drug acts specifically within the cancer cells. This targeted approach significantly reduces the systemic side effects associated with chemotherapy and improves therapeutic outcomes. FA-mediated delivery also benefits from the dual functionality of QDs, enabling both therapy and imaging. The fluorescent properties of QDs allow real-time tracking of the drug delivery process, providing valuable feedback on the distribution and targeting efficiency. However, the stability of folate-QD conjugates in circulation and their potential for immune recognition are challenges that need to be addressed to fully realize the potential of this strategy [54].

4.2.3 RNAi Molecules-Mediated Delivery

RNAi-mediated delivery involves the use of small interfering RNA (siRNA) or microRNA (miRNA) to silence specific genes. QDs can be used to deliver these RNA molecules to target cells, where they induce gene silencing and therapeutic effects. For example, QDs conjugated with siRNAs targeting oncogenes can specifically deliver their payload to cancer cells, reducing tumor growth and providing a high degree of specificity and efficiency in gene therapy [61].

QDs functionalized with siRNA can be designed to target specific cancer cells through receptor-mediated endocytosis. Once inside the cells, the QDs release the siRNA, which then binds to the target mRNA and induces its degradation. This process effectively silences genes, inhibiting the growth and proliferation of cancer cells [65]. Studies have demonstrated the effectiveness of this approach in various cancer models, where QDs have been used to deliver siRNAs targeting oncogenes such as KRAS and BCL-2, resulting in significant tumor reduction [66].

The use of QDs for RNAi delivery also allows real-time tracking and imaging of the delivery process. This capability is crucial for monitoring the efficiency and distribution of the siRNA, ensuring that it reaches the target cells and exerts its therapeutic effect. However, the stability of the siRNA–QD conjugates in circulation and the potential for off-target effects are challenges that need to be addressed to optimize this delivery system for clinical applications.

4.2.4 Glucose-Mediated Delivery

Glucose-mediated delivery exploits the increased glucose uptake in cancer cells. QDs functionalized with glucose or glucose analogues can target these metabolically active cells, ensuring that the drug is delivered specifically to cancerous tissues. This method has been demonstrated with QDs linked to glucose and loaded with anticancer drugs, resulting in enhanced uptake by cancer cells and improved therapeutic outcomes [62]. For example, glucose-conjugated QDs have been used to deliver drugs to cancer cells that exhibit high glucose uptake. The QDs bind to glucose transporters on the cancer cells, leading to receptor-mediated endocytosis and internalization of the QDs. Once inside the cells, the drug is released, ensuring that it acts specifically within the cancer cells. This targeted approach reduces the systemic side effects associated with chemotherapy and improves therapeutic outcomes. A unique study introduced glucose-modified carbon QDs with 2,2,6,6-tetramethylpiperidinooxy (TEMPO) for targeted bimodal MR and optical imaging of tumor cells. The compound CQD-TEMPO-Glu has low cytotoxicity, good biocompatibility, green fluorescence emission, and high T1 relaxivity, enhancing its cellular uptake in cancer cells through GLUT-mediated endocytosis. These properties position CQD-TEMPO-Glu as a promising dual-modal contrast agent for cancer imaging applications [67].

5 Strategies for Mitigating Quantum Dot Toxicity

While QDs are exceptionally versatile tools and their usage is rapidly growing, one cannot neglect the toxicity concerns that accompany them. Since QDs are often made up of heavy metals, their degradation may result in the release of free metal ions [68]. In addition to this enormous risk, they also have a reactive surface, which may lead to cell damage, inflammation or, in worse scenarios, organ failure or cancer. Thus, reducing their toxicity becomes crucial for their successful integration into any theragnostic application. Strategies such as encapsulation, surface coating, and doping have been employed to address these concerns, while factors such as dosage reduction, size modulation, and surface charge manipulation are being studied to understand how they affect the toxicity caused by QDs.

QDs can cause toxicity at multiple biological levels. At the cellular level, QDs often cause toxicity by entering cells via endocytosis. Endocytosis involves the uptake of materials through the use of endosomes, which are lipid vesicles that detach from the cell membrane to transport these materials into the cell. As endosomes have an acidic pH, it is very likely that the QDs within them degrade and release free metal ions. This can progress and lead to damage to cellular organelles and their functions. In addition to affecting organelles, QDs can affect the cell membrane and the genetic material contained within the nucleus. Once the QDs are in contact with the cell membrane, they can easily affect the polyunsaturated fatty acids present in the lipid bilayer by oxidizing their double bonds. This leads to disturbances in their permeability, fluidity, and ionic distribution [3]. QDs are also capable of affecting important organelles, such as mitochondria and the endoplasmic reticulum (ER), by interfering with their morphology. These interferences mostly take place due to the formation of ROS, whose excess accumulation causes oxidative stress in the cell. Upon exposure to CdS QDs at low concentrations, the mitochondria-mediated intrinsic apoptotic pathway in cells can be triggered even without damage to mitochondrial DNA. One of the reasons for the activation of this cell death pathway may be the oxidative stress caused by the ROS generated by the QDs [69]. Similarly, the morphology of the ER can also change after exposure to QDs. Yu et al., 2015 reported that the expression of proteins related to ER stress increased after contact with TiO_2, which is a metal oxide that is often used in the production of QDs [70]. The detection of ER stress is essential, as it is considered one of the first signs of QD cytotoxicity. Another aspect of QD toxicity can be damage to the genetic material of the cell. Like the ER, TiO_2 can also affect the nucleus by causing chromosomal damage and altering gene expression [71]. This process has the potential to induce apoptosis. Smaller green QDs and anionic QDs preferentially accumulate in the nuclei and nucleoli of living cells, which can also cause damage to DNA and modify the process of cell differentiation.

QDs also have the potential to affect organs that have high blood flow, such as the liver, spleen, and kidneys [72]. Mostly, QDs affect organs by causing inflammation via the generation of ROS. The liver is frequently affected by QD toxicity, as it is responsible for QD metabolism and excretion. When the toxicity of Cd-based QDs was studied in a mouse model, inflammation of the liver or hepatitis was detected. Although the exact pathways are still unclear, this inflammation could have been caused by the excess production of ROS, the release of free Cd^{+} ions and/or an increase in the levels of intracellular Ca^{2+} ions [73]. In some cases, prolonged inflammation of the liver can lead to liver damage [74]. In addition to the liver, the kidneys are also affected by QD toxicity. Since they are directly responsible for the excretion of toxic materials from a biological system, there is a high risk of QD accumulation in the kidneys. QDs often interfere with glutathione (GSH) levels within kidneys, which can cause issues with their normal function, as GSH plays an important role in the elimination of free radicals. Moreover, QDs can also adversely affect kidney cells by disturbing their redox balance and can contribute to their apoptosis via the production of excessive ROS [75]. However, Cd-QD toxicity in the liver and kidneys has been shown to be reversible and time-dependent in a study by Wang et al., 2016 [76]. The pulmonary toxicity of QDs, on the other hand, has long-lasting effects, as it does not resolve as quickly as renal or hepatic QD toxicity does [77]. Upon studying mice exposed to QDs for 24 h, Jacobsen et al., 2009 reported that the mice had acute pulmonary inflammation alongside edema and hepatic necrosis [78]. Studies have also established that the majority of nervous system damage occurs through inflammation and oxidative stress, which is alarming, as the brain is particularly susceptible to oxidative stress because of its high oxygen consumption, lipid-rich content, and relatively low antioxidant levels. Moreover, the size of QDs allows them to access the blood–brain barrier (BBB), which is a tightly locked layer of endothelial cells that is semipermeable in nature and strictly regulates the entry and exit of solutes and particles into the brain. Since most neurons do not proliferate, their death due to intense damage can have serious implications. An experiment conducted by Wu et al., 2015 revealed that exposure to CdTe QDs in rat neurons increased ROS and intracellular Ca^{2+} levels, leading to neuronal apoptosis and even death in certain cases [79]. Similarly, Hopkins et al., 2014 reported that QDs can reach the brain through the olfactory bulb when the QDs are aerosolized and are taken up by mice through inhalation. This exposure results in brain inflammation and changes in glucose metabolism pathways, both of which cause damage to the brain [80]. Furthermore, research has also indicated that QDs preferentially accumulate in crucial areas of the brain, such as the corpus callosum and the hippocampus, which increases the vulnerability of these regions to damage [81]. In addition to these organs, QDs can also interfere with the function of reproductive organs or gonads and the spleen, which can be exposed to QDs during bioimaging (Fig. 6).

Thus, effective toxicity mitigation of QDs is needed and requires a multifaceted approach involving encapsulation, surface coating, and doping. Encapsulation techniques involve wrapping a layer of nontoxic material over the outer shell of the QDs, which plays a vital role in reducing their toxicity [82]. Polymer encapsulation via synthetic polymers can effectively increase the biocompatibility of QDs but can often face challenges in terms of their synthesis and stability [83]. Alternatively, encapsulation using green materials such as proteins, biopolymers and lipids can provide more straightforward fabrication processes and improved biocompatibility. Silica encapsulation, on the other hand, is a technique that has been used in abundance, but the interactions of nanonized silica with the human body are still not fully understood. Hybrid encapsulation techniques use a combination of different materials where the benefits of each component can be leveraged to mask the drawbacks of another, providing a versatile approach for QD stabilization and toxicity mitigation. The surface coating of QDs is another critical strategy for reducing toxicity. Here, the surface of the QDs is modified to hold ligands of choice and reduce the amount of QD degradation and ROS generation. Polymer coatings, much like polymer encapsulants, are useful in shielding core QDs from the biological environment. They can, however, lead to liver toxicity, sometimes due to the body's response to synthetic polymers. Therefore, exploring alternative coating materials, such as natural biomolecules, can offer safer and more biocompatible solutions. Doping is another technique that employs the addition of other atoms or compounds to the QD structure, which decreases the amount of toxic metals that make it up. Doping not only decreases the release of heavy metal ions but also mitigates the generation of ROS, thereby addressing two critical aspects of QD toxicity at once.

While QD toxicity can be mitigated via encapsulation, surface coating and doping, multiple factors influence QD toxicity. Reducing the dosage of QDs is likely the simplest method for reducing their cytotoxic effects. Regardless of the type of encapsulation layer or surface coating around the QD, its toxicity is always dependent on its concentration. Properly regulating their dosage may also single-handedly resolve the issue of QD toxicity. Moreover, the size of the QDs also plays a significant role in their biological interactions. Reducing the size of QDs can increase their

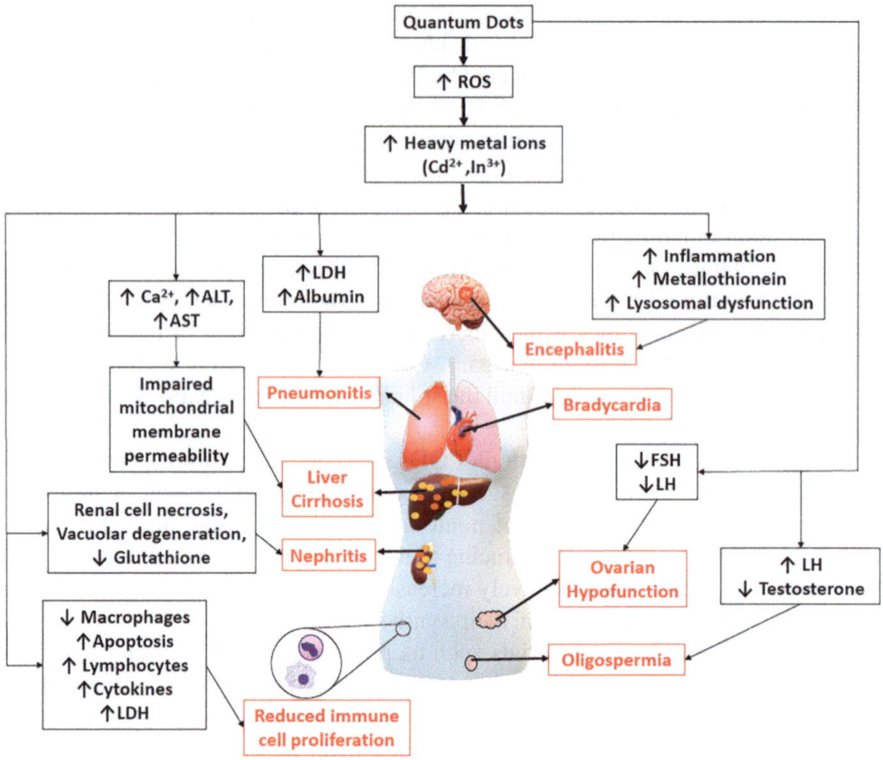

Fig. 6 Schematic representation of the toxic effects of quantum dots (QDs) on various organs in the body. While QDs are being used in different biomedical applications, their toxicity is still a major concern for many. At the cellular level, QDs can enter cells via endosomes and interfere with the functioning of intracellular structures such as mitochondria, the endoplasmic reticulum and DNA. These disturbances occur because QDs are often made up of heavy metals such as cadmium and release free ions upon degradation. Moreover, QDs also generate ROS during photoluminescence. In addition to causing damage to cellular organelles, QDs can also cause damage at the organ level. The liver and kidneys are affected mostly by QD toxicity, as these organs are affected by QD metabolism and excretion. Owing to their small size, QDs can also reach the brain by crossing the blood–brain barrier and inducing oxidative stress within the brain. In addition to these organs, QDs can also travel to reproductive organs and the spleen, where they can cause toxicity. Thus, it is crucial to mitigate the toxicity of QDs before their integration into any biomedical application. Reprinted with permission from [1]

cellular uptake and facilitate efficient targeting to specific tissues or cells. Smaller QDs are more likely to evade the reticuloendothelial system, reduce accumulation in nontarget organs and minimize systemic toxicity. Conversely, there is also a probability that smaller QDs accumulate within cells, as their smaller size allows for better cellular uptake. Additionally, the surface charge of QDs influences their biodistribution, cellular uptake, and excretion. By carefully controlling the surface charge, it is possible to mitigate toxicity further and improve the targeting efficiency of QDs.

5.1 Encapsulation of Quantum Dots

Encapsulating a QD implies the introduction of an additional layer on top of the outer shell of the QD. This outer layer can consist of various materials, such as polymers, lipids, proteins, silica, or combinations thereof. An encapsulated QD is shown to have less toxicity, as this additional layer decreases the number of interactions the QD has with the cell and the biological system. This also ensures that the QDs do not degrade within the cell and are toxic because of the release of free heavy metal ions and/or ROS. In some cases, the encapsulation layer increases the size of the QD, which in turn reduces its accumulation within certain organs, such as the liver and spleen. Similarly, it can affect the surface charge of the QD and alter the organs in which it is deposited. On the other hand, the dosage limits of encapsulated QDs are much higher than those of free QDs.

The encapsulation layers of a QD can be classified into four major categories: green, synthetic, silica and hybrid encapsulants. Green encapsulation of QDs involves the use of materials sourced from biological samples such as animals, plants, bacteria, and fungi. The utilization of such materials is gradually increasing because of their enhanced biocompatibility and negligible toxicity. Conversely, synthetic or man-made materials comprise laboratory-prepared polymers or lipids that are biodegradable and are mostly biocompatible in nature. Silica is an inorganic compound that is especially renowned for its inert nature. However, silica-encapsulated QDs have still not been introduced on the market because of the limited knowledge available on the interactions that nanonized silica has with various biological systems. Hybrid encapsulants involve the use of more than one type of encapsulation material. This is a unique technique, as it can be used to formulate extremely efficient encapsulation layers by using multiple materials to resolve possible drawbacks.

5.1.1 Green Encapsulation

Green encapsulation consists of extracting encapsulation materials from biological sources such as plants, fungi, and bacteria. These constituents can be proteins, polymers, lipids, or bioactive compounds (Fig. 7d). While much work has not been done on encapsulating QDs with green materials, many studies have attempted to successfully synthesize plant-based QDs, especially carbon dots. The use of green materials for toxicity mitigation is a broadly exploited idea, as these materials are sourced from the very system in which they reduce toxicity, making them highly biocompatible. Additionally, these products are easily excreted from a biological system, as such products are ingested regularly in the form of food. Green materials can also be obtained from numerous sources, increasing novelty and customization. In addition to reducing toxicity, green materials can also provide additional benefits, such as antioxidant and anti-inflammatory properties. For example, polyphenolic compounds from *Aloe vera* have been used for encapsulating ZnS:Mn QDs. While

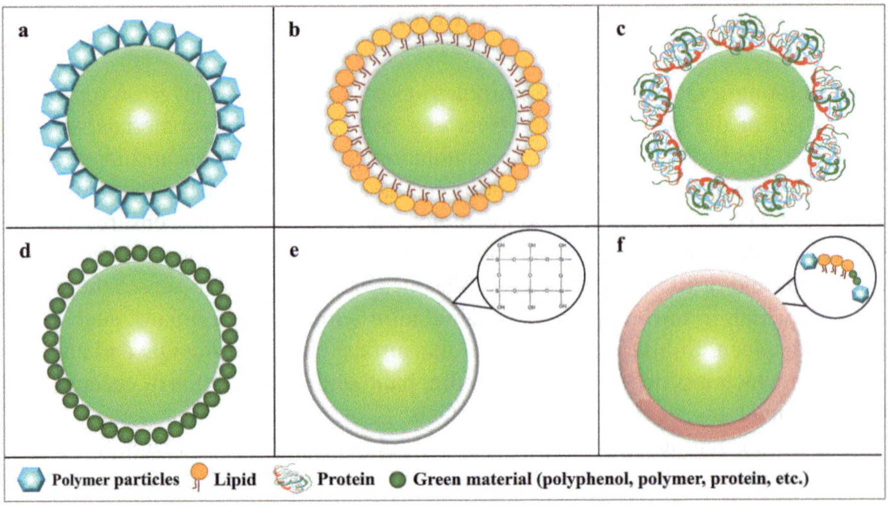

Fig. 7 Diagrammatic representation of quantum dots (QDs) with different types of encapsulations. (**a**) Polymer encapsulation can be achieved using synthetic or bio-polymers. They significantly reduce toxicity of QDs but can be problematic in terms of synthesis, (**b**) Lipid encapsulation of QDs increases biocompatibility but it does so at the cost of faster fluorescence lifetimes, (**c**) QDs have not been encapsulated in protein layers frequently but that also brings the question of their viability. During future investigation, studies must ensure that the proteins involved do not aggravate toxicity themselves, (**d**) Green materials are increasing becoming a popular choice for entrapping QDs due to their added benefits. They can be extracted from plants, fungi, animals, bacteria, etc., and can be either proteins, polymers or polyphenolic compounds, (**e**) Silica is widely used for encapsulating QDs but is mostly used in electronic applications. In terms of toxicity, silica encapsulated-QDs have not been put through any human trials as we still lack the full understanding of the effects silica has inside the body, (**f**) Hybrid encapsulation, though an infrequently used mode of toxicity mitigation, has great potential. It can not only increase the efficiency of the encapsulation layer but also target toxicity reduction in multiple ways due to the added benefits of numerous encapsulation materials. The figure was made using Microsoft® PowerPoint® Version 2406 Build 17726.20126 (Microsoft®, Redmond, WA, USA)

unencapsulated QDs showed cytotoxicity at concentrations higher than 6.25 μg/ml, the *Aloe vera* extract-encapsulated QDs were much less toxic. They only exhibited toxicity at concentrations higher than 25 g/ml, leading them to be four times less toxic than regular QDs [84]. This large difference in toxicity highlights the efficacy of polyphenolics such as flavonoids and phenolic compounds as encapsulation materials. Flavonoids and phenolic compounds are secondary metabolites that are specific to plants and are produced by the plant body under stress conditions.

5.1.1.1 Green Encapsulation Using Proteins

Protein encapsulation consists of extracting proteins from biosources and then utilizing them to encapsulate QDs (Fig. 7c). For example, a study successfully encapsulated graphene QDs or GQD-iron complexes within ferritin nanocages. Ferritin is a cytosolic protein that acts as an iron reservoir in cells and is crucial for the maintenance of iron homeostasis. Since GQDs are inherently more biocompatible and ferritin is a biologically sourced protein, the cytotoxicity levels of such nanoparticles are much lower than those of regular CdSe QDs. Photoluminescence studies revealed that there was no significant quenching in the fluorescence intensity of the GQDs, confirming that protein encapsulation did not affect the competence of the GQDs [85]. However, ferritin is a protein that is suspected to aid in the proliferation of breast cancer cells, making it a dangerous candidate for encapsulating QDs that can be used for purposes involving this specific type of cancer. This study, hence, conducted cytotoxicity studies on two breast cancer cell lines—MDA-MB-231 and 3T3. The cytotoxicity generated by such QDs is a sum of the toxicity produced by each component of the QD. While ferritin did not promote the proliferation of MDA-MB-231 cells and significantly decreased their growth rate, the release of iron ions during the degradation of the nanocages at lower pH led to concern about iron toxicity. Iron toxicity is especially dangerous, as it results in the production of ROS whose overproduction has the potential to push normal cells into cancer. Additionally, other studies have confirmed the role of ferritin in promoting cancer progression, thereby sustaining concerns regarding its use in cancer-related QD research [86].

A better model for protein encapsulation is henceforth a protein that is not involved integrally in maintaining cellular homeostasis but is also biologically sourced. This is because any change in its concentration in a cell due to its introduction via encapsulation would not directly affect the metabolism of the cell. Yu Lee et al., 2014 created CdSe and CIS/ZnS QDs and then covered them with a layer of human serum albumin (HSA) crosslinked with poly(ethylene glycol) or PEG [87]. HSA is a plasma protein, whereas poly(ethylene glycol) is a biocompatible polymer. The cytotoxicity of these encapsulated QDs was tested in three cell lines: HeLa cells (human epithelial carcinoma cells), PC3 cells (human prostate cancer cells), and A549 cells (human adenocarcinomic alveolar basal epithelial cells). As expected, a lower cell viability was observed for the nanoparticles containing a CdSe core than for the CIS/ZnS-cored nanoparticles. However, the protein encapsulation in this case was futile, as the difference in the QD core was the cause of the lower degree of toxicity.

5.1.1.2 Green Encapsulation Via Polymers

Biologically sourced polymers such as chitosan and gelatin are some of the most frequently used polymers for the green polymer encapsulation of QDs. They are wrapped around the outer surface to limit the interactions between the environment and the QDs (Fig. 7a). This ensures delayed degradation and less exposure of the QDs to the cells. Since these are polymers that have been extracted from living systems, they are very biocompatible and, like polyphenols, can offer additional benefits. Chitosan is a biopolymer made up of multiple D-glucosamine monomers. It is naturally found in the shells of invertebrates such as crabs and lobsters. As it is an edible material, its excretion from the body is also easily possible via feces. Like chitosan, gelatin is also a biopolymer that can be extracted from algae and is edible.

The addition of such polymers can make QDs extremely versatile, as they can potentially carry drugs across a system while removing the toxicity caused by the QDs. This allows the encapsulation layer to prevent cell damage in more than one way. Chitosan carries multiple protonated amino groups that can be easily functionalized to carry drugs. These drug-loaded encapsulation layers prevent the release of heavy metal ions from the QDs and/or quench ROS as they are produced by the QDs. A study by Tan et al., 2007 reported that chitosan-encapsulated QDs exhibited improved cytotoxicity [88]. Like chitosan, gelatin can also reduce the cytotoxicity of QDs by encapsulating them. Chen et al., 2013 reported that the cell viability of gelatin-encapsulated QDs was drastically greater than that of unencapsulated QDs. While gelatin-encapsulated QDs were not toxic to the cells until a concentration of 5 mg/ml was reached, free QDs were toxic at much lower concentrations [89].

5.1.2 Synthetic Encapsulation

Synthetic encapsulants of QDs constitute materials that are man-made or produced in laboratories. These materials can be designed to be inert, biocompatible, biodegradable and nontoxic to biological systems. However, their production is expensive, and in some cases, they may trigger allergic reactions despite their inert nature. However, they are popularly used for QD encapsulation, as they have been shown to successfully prevent the degradation of QDs. Additionally, the production of such materials is less labor intensive than extraction and the subsequent testing of green materials. The production of synthetic polymers and lipids is also more time-efficient.

5.1.2.1 Synthetic Encapsulation Using Polymers

Polymer encapsulation is a technique that involves entrapping QDs in small polymer nanoparticles or microspheres. Multiple methods, such as seeded emulsion polymerization and self-polymerization, can be employed to successfully encapsulate QDs. To address toxicity, polymers such as poly(ethylene glycol) (PEG) and polyethyleneimine (PEI) are frequently used. These synthetic polymers have been used for biomedical purposes repeatedly because of their biocompatibility and biodegradability. Although there are few concerns involving this type of encapsulation, Sheng et al., 2005 published a study in which they described the complexities involved in the entrapment of QDs in polymeric microspheres. They detail its instability and call for better modified protocols for improved loading efficiencies of QDs within the microspheres [90]. This may contribute to increasing toxicity because untimely degradation of polymer-encapsulated QDs can lead to cell damage and heavy metal accumulation.

However, stable polymer encapsulants result in good cell viability. Arugete et al., 2010 studied QDs surrounded by polymers and placed them in a bacteria-exposed environment. Neither did the bacterial cells die upon exposure to the QDs, but the stability of the QDs was also not affected when they were exposed to the bacteria [36]. This preliminary investigation of bacterial cells supports the reasoning that this method of encapsulation does not pose a significant cytotoxicity risk to eukaryotic cells. Similarly, Romoser et al., 2011 synthesized PEG-encapsulated QDs, which showed no toxicity, as the encapsulation layer prevented the contact of the QDs with the cells and, consequently, their cellular uptake [91]. Thus, it can be concluded that polymer encapsulation of QDs is a viable option for mitigating toxicity; however, the questions of their proper preparation and stability remain.

5.1.2.2 Synthetic Encapsulation Using Lipids

Lipid encapsulation involves entrapment of QDs in lipid bilayers or lipid molecules to reduce the toxicity of the QDs (Fig. 7b). For encapsulation, the charge of a lipid becomes an essential criterion to consider, as the interactions between the proteoglycans present on the cellular membranes and the liposomes entrapping the QDs are mostly electrostatic in nature [92]. This makes cationic lipids such as 1,2-dioleyl-sn-glycero-3-phophocholine (DOPC) and 1,2-dioleyl-sn-glycero-3-phopho-L-serine (DOPS) popular choices because they not only increase the cellular uptake of QDs but also structurally resemble phospholipids of the lipid bilayer making up the cell membrane.

As lipids are an integral part of cellular membranes, their encapsulation can increase the internalization of QDs [93]. A study conducted by Brunn and Hille, 2019 revealed that free QDs were not capable of proper internalization by their target cells but rather accumulated in the surrounding environment. The encapsulated QDs, on the other hand, show strong luminescence at the cell surface, which gradually decreases and then quenches completely in the cytosol, thereby preventing any possible accumulation of the QDs in the cellular environment [92]. This is especially beneficial, as the QDs are being used up and excreted out of the system instead of accumulating within the biological system.

Although lipid encapsulation of QDs is highly favorable, it has multiple concerns at the same time. The interactions between liposomes and the cell membrane are not specific in nature. This can lead to the uptake of QDs by cells that have not been targeted. Such an interaction can very likely take place at a location much further away from the target cell or organ, as lipid encapsulation often increases the water solubility of QDs [94], enabling them to travel further distances via the bloodstream. Thus, careful consideration must be given when deciding on such an encapsulation layer and its mode of delivery into the body. Unlike some polymer and protein encapsulants, lipid-encapsulated QDs do not maintain their efficiency. Generalov et al., 2011 studied the fluorescence lifetimes of such entrapped QDs. As the QDs were irradiated, their fluorescence intensities increased dramatically, which ultimately caused them to be extinguished faster. However, this is an additional advantage in terms of lowering toxicity, as it results in the generation of fewer free radicals [95]. This phenomenon diminishes concerns surrounding the build-up of ROS in cells. As the QDs are quenched faster, their degradation and resulting excretion from the biological system also occur faster. Thus, for toxicity mitigation, lipid encapsulation is a very good method, as it efficiently addresses this issue despite affecting the functionality of the QDs.

5.1.3 Silica Encapsulation

Silica encapsulation uses an inorganic molecule called silica or SiO_2 to encapsulate QDs. Even though it is not an organic molecule, silica is relatively biocompatible. However, its biocompatibility has long been a concern because of its structure and varying sizes. Like green encapsulation, silica is mostly used for synthesizing silica-based QDs, where the outer shell is made up of silica. It is relatively less toxic than CdSe/ZnS core–shell QDs because one of the layers is made up of a nontoxic material. With respect to encapsulation, mesoporous silica (MS) is mostly used. This is because of its greater biocompatibility and stability than those of dense silica and most synthetic polymers (Fig. 7e) [96]. MS is known for its ability to reach nanoscale sizes, high surface area, vast tunability and structures with large pore volumes. They are often incorporated into encapsulating QDs in the form of MS nanoparticles [97]. Nonetheless, studies have shown that these compounds can interfere with cellular metabolism [98].

While MS has a plethora of benefits, they remain a hazard, as we still do not completely understand all their interactions within a biological system. It has been established that their size, shape and surface charge play integral roles in their biocompatibility and cellular uptake. The size of the MS nanoparticles is dependent on their dosage. However, as their interactions with cells are not completely understood, it cannot be predicted what complications may arise. The shape of an MS nanoparticle also contributes greatly to its internalization. While spherical silica nanoparticles are taken up easily, rod-shaped particles remain in the bloodstream for longer periods of time [98]. However, there is ambiguity in this assumption because a study by Huang et al., 2010, proved that the inverse was also true and that rod-shaped silica-encapsulated QDs were taken up better than spherical QDs [99]. Like the shape of MS, its surface charge influences its cellular uptake. Souris et al., 2010 published a report in which different charges of silica were used for encapsulating QDs. Negatively charged silica-encapsulated QDs accumulate in the liver sinusoids, but positively charged silica-encapsulated QDs are eliminated from the body without any toxicity issues [96].

5.1.4 Hybrid Encapsulation

Although hybrid encapsulation has not been explored extensively, it involves the use of multiple materials for encapsulating QDs. This can encompass the usage of polymers, lipids, polyphenolics, proteins, and/or silica (Fig. 7f). To date, work has focused only on polymer–lipid hybrid encapsulations and polymer–silica hybrid encapsulations. Kim et al., 2010 encapsulated QDs in a PEG-PCDA (10,12 pentacosadiynoic acid) layer. This improved the cytotoxicity of the QDs while increasing their water solubility. One of the reasons for the increased water solubility was the stability imparted by the polymer moiety [100]. A similar study was conducted by Hu and Gao, 2010, who used an encapsulation of PE-PEG and SiO_2 instead. The incorporation of the amphiphilic polymer into the encapsulation ensures greater stability of the QDs under acidic conditions [101], providing prospects for bioimaging in the GIT tract without QD degradation or toxicity.

Hybrid encapsulation, although less investigated, has great potential for toxicity mitigation. With careful selection of materials, one can resolve the drawbacks of a certain type of encapsulating material with another. For example, the drawbacks of encapsulation involving proteins, which may alter cellular metabolism, can be mitigated by the conjugation of green materials or polymers, which may reduce the effects of protein and QD cytotoxicity. Similarly, synthetic polymeric encapsulation can be improved by incorporating lipids, proteins and/or silica. The water solubility of a QD can also be increased exponentially by the combination of amphiphilic copolymers and lipids in the encapsulation layer.

5.2 *Surface Coating of Quantum Dots*

Surface coating involves modifying the outer shell of the QD to bear the ligand of choice. They can be conjugated with drugs or other materials and can be made multifunctional. Surface coating and encapsulation are two terms that can be easily confused with each other. However, in the case of encapsulation, the complete quantum dot is entrapped or wrapped within an additional layer. Unlike surface coatings, encapsulants do not require any chemical modifications to the surface structure of the outer shell of the QDs.

Surface modifications for the incorporation of coatings can be achieved in two ways: (1) replacing the ligands present in the outer shell with new ligands or (2) conjugating the new ligands to the existing ligands. Either way, surface coatings are a reliable method for reducing the cytotoxicity of QDs, as they can prevent the release of free ions from the QDs and increase their stability. There are five different methods or materials that can be used to modify the outer shell of a QD. These materials include polymers, multipurpose ligands or multiligands, zwitterions, peptides and inorganic materials such as silica (Fig. 8).

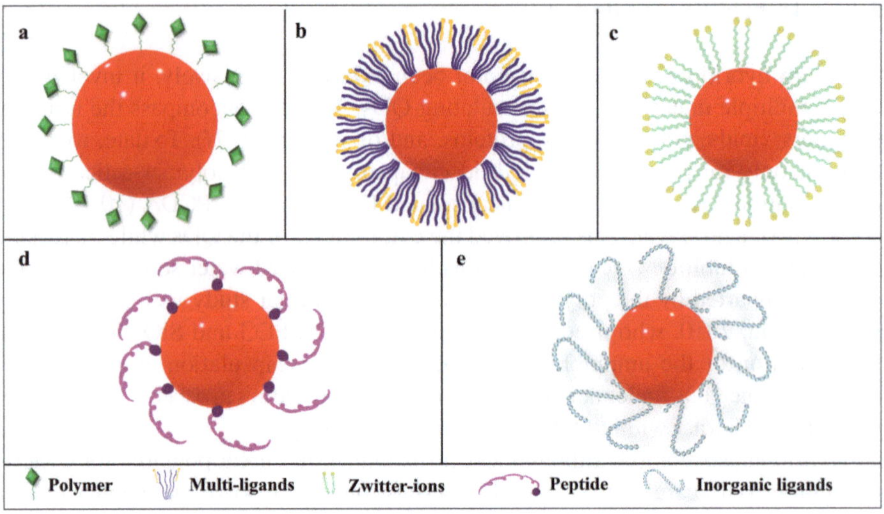

Fig. 8 Schematic representation of various surface coatings that can be applied onto quantum dot (QD) surfaces. (**a**) Amphiphilic and multidentate polymers can be applied onto QD surfaces to get polymer coatings. These are more widely used than polymer encapsulations but run the risk of hepatitis or liver inflammation in some cases, (**b**) Multi-ligands are functionalised ligands on QDs which have more than one function. These ligands can be peptides, polymers, zwitter-ions or silica and achieve biocompatibility whilst targeting multiple objectives, (**c**) Zwitter-ions are ligands which contain a positively charged moiety as well as a negatively charged one. They use surface charge to reduce QD toxicity and can be easily converted into multi-ligands, (**d**) Peptide coatings also use surface charge for toxicity mitigation and are a more biocompatible method than protein encapsulation, (**e**) Inorganic coatings mostly make use of silica due to its inert nature. In some cases, they are shown to have even lesser toxicity than organic coatings. The diagram was made using Microsoft® PowerPoint® Version 2406 Build 17726.20126 (Microsoft®, Redmond, WA, USA)

5.2.1 Polymer Coating

Polymer coating is a technique in which synthetic polymers are used to coat a QD to reduce its toxicity and increase its functionality. Compared with polymer encapsulation, polymer coating is a much more established and widely used technique. Here, two types of polymers are mostly used—multidentate and amphiphilic copolymers (Fig. 8a). Compared with monodentate polymers, which can form only one bond per functional group, multidentate polymers such as thiols and oligomeric phosphines offer better bonding stability because of their ability to form a greater number of bonds. Alternatively, amphiphilic copolymers provide better stability to the QD in different media [102]. Nonetheless, both of these polymer types reduce the degradation of QDs within the body and aid in their excretion by increasing their solubility.

While polymer coatings can mitigate the toxicity of QDs, there are concerns regarding their long-term use in applications such as bioimaging owing to the potential for intraendosomal degradation. This degradation occurs within endosomes due to factors such as acidic pH, which not only results in the disruption of normal cellular functions but also diminishes the efficacy of the QDs [102]. Another study conducted by Smith et al., 2012 investigated Tri-*n*-octylphosphine oxide, poly(maleic anhydride-*alt*-1-tetradecene) copolymer (TOPO-MAT)-coated CdSe/ZnS QDs and their effects on two liver cell lines—PHL (primary human liver) cells and HepG2 cells. The HepG2 cells effectively internalized the QDs into the subcellular vesicles, but only some cells in PHL displayed the same positive results. In both cases, however, stress and inflammatory marker expression were increased. This finding proved that while there is no apparent cytotoxicity at that concentration, it is very likely that these encapsulated QDs cause hepatitis if they remain in the system for too long [103].

5.2.2 Multiligand Coating

Multiligand or multipurpose ligands are any type of ligand that has been attached to the outer surface of a QD to not only mitigate toxicity but also deliver materials such as drugs. They can be made up of lipids, polymers, proteins or a combination of them conjugated together (Fig. 8b). Versatility is a key feature of this type of surface coating, as the ligands being used can be self-synthesized or customized according to their function. This makes toxicity studies extremely crucial, as the ligands may or may not always confer biocompatibility.

Studies have reported improved cytotoxicity of QDs after multiligand coating. Pan and Feng, 2008 constructed QDs with a polymer coating conjugated with vitamin E [54]. Here, the QDs were functionalized to target breast cancer cells. In spite of this, they showed lower cytotoxicity in normal cells, making them successful in their multifunctionality. Similarly, Wang et al., 2016 made liposomic acid-amine (LA)-poly(isobutylene-*alt*-maleic anhydride) (PIMA) ligands, which were then attached to the outer surface of the QDs [104]. These materials were designed to provide increased cellular uptake, which they successfully accomplished, making the use of multiple ligands a very useful technique for mitigating toxicity.

5.2.3 Zwitter-Ion Coating

A zwitterion is a molecule that contains a positive charge as well as a negative charge. Its ionic state is dependent on its surrounding conditions, especially pH. Although most proteins and peptides exhibit this characteristic, many molecules can also be made into zwitterions. Coatings made out of these molecules can be employed not only to target QD toxicity but also to target organs and cells (Fig. 8c). Like multiligand coatings, zwitterion coatings can also be produced in the laboratory. However, in this case, these coatings engage surface charges to mitigate toxicity instead of having multiple functions or adding drugs.

To increase hydrophilicity and biocompatibility, imidazole and zwitterion-based ligands were developed for colloidal stability. One of the reasons for their improved stability could be attributed to zwitterions, which increase the interactions between the media and the QDs [105]. Susumu et al., 2011 created similar zwitterion ligands, which also contributed to increasing biocompatibility [106]. A study also tested different zwitterion ligands with varying surface charge densities and reported that the higher the charge density was, the lower the cytotoxicity was. They also concluded that the positively charged zwitterions and the negatively charged cell membrane interacted more, which resulted in greater uptake and thus toxicity. As expected, the inverse was true in the case of negatively charged zwitterion ligands [107]. Zwitter ions are thus multifaceted ligands that can reduce toxicity in two ways: (1) their chemical composition and (2) their surface charge and resulting interactions.

5.2.4 Peptide Coating

Peptides are short chains of amino acids that come together and arrange themselves in chains to form proteins. These singular chains are used as coating ligands in peptide coatings (Fig. 8d). As mentioned previously, peptides also have zwitterion qualities. This implies that within a body, the peptide coating may also use surface charge to reduce toxicity. Wei et al., 2009 reported that in the case of peptide coatings, cellular internalization increased as the surface charge increased. However, they also highlighted that as the conjugation method or the method via which peptides are added onto the QD changes, their properties also change. For example, when the ligands were conjugated via direct ligand exchange, the colloidal stability of the QDs decreased [108].

In addition to using biosourced peptides, one can also synthesize them in laboratories. An experiment by Xu et al., 2012 revealed a new peptide (phytochelatin) that increased the water solubility of QDs. Upon testing on live zebrafish, these QDs also did not show any cytotoxicity [109], establishing that peptide coating is a more stable and reliable method of using peptides for reducing toxicity. Compared with protein encapsulation, peptide coating is more biocompatible, as the peptide chains do not contribute to cell toxicity themselves.

5.2.5 Inorganic Coating

Inorganic coating of QDs mostly involves the use of silica particles. These particles are then used to entrap QDs to reduce QD toxicity (Fig. 8e). In contrast to silica encapsulation, silica coating is a much less toxic method of incorporating silica. These coatings are larger in size than MS and thus do not cause as many issues within the body. Moreover, their sizes remain relatively constant, ensuring that these QDs do not accumulate in various regions of the body. In some cases, QDs with such coatings have been observed to be more biocompatible and have lower toxicity levels than organic coatings [110]. The inert nature of silica could be one of the reasons for such an observation.

Since silica particles are conjugated with the QD surface in a surface coating, they are less likely to degrade and interfere with cellular metabolism. Silica-coated QDs have also been shown to remain stable through exposure to heat, light and electrolytes while reducing the cytotoxicity of the QDs [111]. These QDs were shown to have diminished cytotoxicity, as the silica coatings controlled the release of heavy metal ions and were also found to be water soluble. These findings prove that silica-coated QDs are stable under different conditions and can be easily integrated into biomedical applications.

5.3 *Doping of Quantum Dots*

QD doping involves the addition of atoms that are not originally a part of the QD layer into the QD shells to enhance their properties or, in some cases, mitigate toxicity. This technique involves the addition of either silica, carbon or any less toxic metal into the QD structure. For example, Liu et al., 2011 created gadolinium (Gd)-doped ZnO QDs for bioimaging purposes. Although these QDs did not cause as much toxicity as free Cd-based QDs did, they were still quite toxic. At a dosage of 5 mM, these QDs had a cell viability of 50%, whereas any material with a cell viability of 80% or more is usually considered safe. Moreover, these Gd-doped ZnO QDs are at risk of release of free Gd^+ ions upon QD degradation [112]. Gd is a rare-earth metal that can deposit in the brain, bones and especially in the kidneys. It can also be retained within the body for longer durations, and this condition is often referred to as the 'gadolinium storage condition', which allows for more Gd^+ ions to cause more issues within the body [113].

A similar study created fluorine (F)- and chlorine (Cl)-doped QDs, which were used as antioxidant agents [114]. Antioxidants are compounds that can use excess ROS and prevent oxidative stress from occurring. Although F- and Cl--based QDs are biocompatible and display good cell viability, there is still potential for possible toxicity. Toxicology studies of fluorine report that, in accordance with its dosage and the total time of its presence within the body, fluorine can be taken up by either soft tissues or hard tissues such as bones and teeth. The majority of fluorine is excreted in the urine, thus increasing the likelihood of its accumulation in the kidneys [115]. On the other hand, chlorine is an agent that is a part of multiple biological warfares and can have chronic effects on the respiratory system when inhaled [116].

Therefore, the use of materials such as silica and carbon is preferable because they do not aggravate the toxicity caused by the QDs. Qian et al., 2014 developed silica-doped carbon QDs, which showed negligible toxicity even though they had been internalized by the cell and had reached the cytoplasm [117]. Moreover, their doping did not have any effect on their functionality, indicating that doping could be used to mitigate toxicity successfully while maintaining the efficacy of the QDs. An experiment conducted by Kumar et al., 2019 involved the preparation of metal-doped carbon QDs for neuroengineering purposes. These CQDs were doped with gold, silver, gallium, tin and zinc, all of which are very less reactive in a biological system. As anticipated, all these QDs did not show any cytotoxicity, except for the Zn-doped CQDs, which could be due to their doping and dosage. However, all of them are safe at a concentration of 0.2 mg/ml, making them suitable for use in neural tissue-related applications [118]. In conclusion, doping QDs with either inert metals or biocompatible materials such as silica and carbon is desirable, and studies have proven their contribution to reducing QD toxicity.

5.4 Factors Affecting Quantum Dot Toxicity

When considering the mitigation of the toxicity of QDs, it is essential to properly understand factors such as dosage limitations, the surface charge of the QDs and their size. The accumulation of different QDs within the body and their resulting toxicity are determined primarily by these criteria. Dosage reduction involves lowering the dosage of QDs that is required for a specific function while not affecting their efficiency. The surface charge of a QD refers to the overall net charge present around the outer surface of the QD, which can be modulated to prevent the accumulation of the QD in specific organs or parts of the cell. The size of the QDs, on the other hand, plays a more direct role in the preferential localization of QDs in various regions of a biological system.

5.4.1 Dosage Reduction

When considering toxicity mitigation in QDs, it is essential to consider their dosage. Dosage involves the concentration of QDs that have been introduced into a body and the safe level of QD concentration that can remain in it for a certain time period. Regardless of the fact that a QD is encapsulated, coated or doped, its concentration can singularly determine the amount of damage it causes. Thus, it becomes necessary to use lower concentrations of QDs when working with them.

The effects of different concentrations of QDs when introduced to cell lines have been studied. If 0–10 μg/ml QDs were given to a cell culture, irrespective of the QD size, the cells would not be affected. However, if the concentration was increased to 10–100 μg/ml, the metabolic activity of the cells decreased [119]. This could be attributed to the higher concentration of QDs present within the cells, which could lead to the generation of ROS and interfere with the normal function of the cell. However, this could decrease the efficiency of QDs, especially in bioimaging. Thus, there is an inevitable need to create better QDs that are more efficient and have greater fluorescence.

5.4.2 Surface Charge

The charge on the surface of a QD is integrally involved in its internalization. In general, cationic materials are taken up better than anionic or neutral particles are taken up better, as the negatively charged cellular membrane interacts better with positively charged compounds. A study by Chinnathambi et al., 2024 revealed that while positively charged coatings resulted in increased cellular uptake, they also did not trigger an immune response. Upon exposure to neutral and negatively charged coatings, immune responses are triggered, and there is increased expression of inflammatory cytokines such as IL-6 and INF-α [120].

In contrast, the preferential accumulation of QDs is also partly dependent on their surface electrochemistry, which includes their surface structure and surface charge. Positively charged or cationic polymers tend to accumulate more in the lung than in the liver, whereas the opposite is true for negatively charged polymers. It is also likely that higher doses of positive QDs may lead to pulmonary embolism due to their accumulation within the lungs (Fig. 9). Pulmonary embolisms block moving materials within the blood vessels surrounding the lungs [121]. Thus, choosing a specific kind of charged surface coating for a QD while keeping in mind its function and target site critically important.

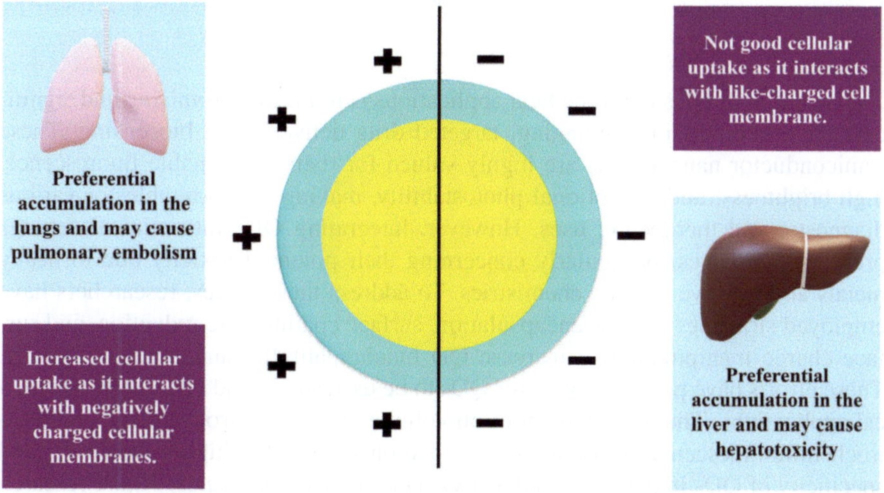

Fig. 9 Effects of surface charge on the internalization of quantum dots (QDs) and their toxicity. Compared with negatively charged or neutral QDs, positively charged QDs are taken up more easily because they interact better with negatively charged cellular membranes. Cationic QDs preferentially accumulate in the lungs, whereas anionic QDs tend to localize in the liver. The figure was created via Microsoft® PowerPoint® Version 2406 Build 17726.20126 (Microsoft®, Redmond, WA, USA)

5.4.3 Size Modulation

Another important criterion to consider when attempting to reduce QD toxicity is size. The size of a QD is measured in terms of its hydrodynamic diameter, which is the diameter of a sphere that experiences the same amount of drag force as itself when it moves across a liquid. In general, the smaller the QD is, the greater the degree of toxicity it causes [3]. This is considered because the contact area of a smaller QD increases, allowing it to cause more damage. However, this may not be true in many cases.

The cellular localization of CdTe QDs is regulated by their size, as red QDs preferentially accumulate in the cytoplasm, whereas green QDs are found in the nucleus. Despite the formation of aggregates, the green QDs could still enter the nucleus, whereas the red QDs could not [122], highlighting their potential toxicity. A similar study revealed that smaller QDs are capable of penetrating barriers such as the blood–brain barrier, placental barrier, and nuclear pore. However, smaller QDs are excreted faster because of their size and ease of movement and thereby do not stay in the system for long enough time to cause much damage. Larger QDs remain in the bloodstream longer and have the ability to cause intense damage. The larger QDs also tend to accumulate more in organs such as the spleen, whereas the smaller QDs settle in the kidneys [3].

6 Conclusion

The potential of QDs in biomedical applications is immense, promising and significant advancements in bioimaging, targeted drug delivery, and biosensing. These semiconductor nanocrystals are highly valued for their size-tunable fluorescence, high brightness, and exceptional photostability, making them suitable for various diagnostic and therapeutic uses. However, integrating QDs into clinical settings presents challenges, particularly concerning their potential toxicity due to heavy metals and reactive surface chemistries. To address these issues, researchers have employed strategies such as encapsulation, surface coating, size reduction, and surface charge manipulation to increase QD biocompatibility and mitigate toxicity. These efforts have paved the way for QDs to be used safely and effectively in medical applications. The development of photoluminescent, electrochemical, and electrochemiluminescence biosensors has demonstrated the high sensitivity and specificity of QDs in detecting and analyzing biological molecules. Future research may focus on creating even more sophisticated biosensors for real-time monitoring of biological processes, enhancing early disease detection and personalized medicine. In targeted drug delivery, the ability of QDs to exploit both passive and active targeting mechanisms opens new possibilities for the precise treatment of diseases such as cancer. Future advancements might include more efficient delivery–release modes and better functionalization techniques to improve targeting accuracy and therapeutic outcomes. Additionally, ongoing efforts to reduce the size and optimize the surface charge of QDs will likely yield more effective and safer drug delivery systems.

In conclusion, while the promise of QDs in biomedical applications is clear, their successful integration into clinical and diagnostic settings will depend on continued innovation in addressing toxicity concerns. Encapsulation, surface coating, size reduction, and surface charge manipulation are critical strategies that have shown significant potential in enhancing the biocompatibility of QDs. Future research should focus on refining these approaches and exploring new methods to ensure the safe and effective use of QDs. By overcoming these challenges, QDs can fully realize their potential to revolutionize bioimaging, targeted drug delivery, and biosensing, ultimately leading to improved disease detection, treatment, and monitoring. The future of QDs in medicine is bright, with ongoing advancements promising for transforming biomedical science and significantly enhancing patient care.

Acknowledgement The authors would like to acknowledge Dayananda Sagar University, Bengaluru, India.

References

1. Biswas, P., Saha, A., Sridhar, B., Patel, A., & Desai, B. M. A. (2023). Quantum dots as functional Nanosystems for enhanced biomedical applications. *Biological Regulators, 38*. https://doi.org/10.23812/j.biol.regul.homeost.agents.20243803.145
2. Iannazzo, D., Espro, C., Celesti, C., Ferlazzo, A., & Neri, G. (2021). Smart biosensors for cancer diagnosis based on graphene quantum dots. *Cancers (Basel), 13*, 3194. https://doi.org/10.3390/cancers13133194
3. Lin, X., & Chen, T. (2023). A review of in vivo toxicity of quantum dots in animal models. *International Journal of Nanomedicine, 18*, 8143–8168. https://doi.org/10.2147/IJN.S434842
4. Wagner, A. M., Knipe, J. M., Orive, G., & Peppas, N. A. (2019). Quantum dots in biomedical applications. *Acta Biomaterialia, 94*, 44–63. https://doi.org/10.1016/j.actbio.2019.05.022
5. Gao, X., Cui, Y., Levenson, R. M., Chung, L. W. K., & Nie, S. (2004). In vivo cancer targeting and imaging with semiconductor quantum dots. *Nature Biotechnology, 22*, 969–976. https://doi.org/10.1038/nbt994
6. Michalet, X., Colyer, R., Antelman, J., Siegmund, O., Tremsin, A., Vallerga, J., & Weiss, S. (2009). Single-quantum dot imaging with a photon counting camera. *Current Pharmaceutical Biotechnology, 10*, 543–557. https://doi.org/10.2174/138920109788922100
7. Liu, X., Braun, G. B., Qin, M., Ruoslahti, E., & Sugahara, K. N. (2017). In vivo cation exchange in quantum dots for tumor-specific imaging. *Nature Communications, 8*, 343. https://doi.org/10.1038/s41467-017-00153-y
8. Mulder, W. J. M., Strijkers, G. J., Nicolay, K., & Griffioen, A. W. (2010). Quantum dots for multimodal molecular imaging of angiogenesis. *Angiogenesis, 13*, 131–134. https://doi.org/10.1007/s10456-010-9177-x
9. Santra S, Dutta D, Walter GA, Moudgil BM. Fluorescent Nanoparticle Probes for Cancer Imaging. Technology in Cancer Research & Treatment. 2005;4(6):593-602. https://doi.org/10.1177/153303460500400603

10. Popovtzer, R., Agrawal, A., Kotov, N. A., Popovtzer, A., Balter, J., Carey, T. E., & Kopelman, R. (2008). Targeted gold nanoparticles enable molecular CT imaging of cancer. *Nano Letters, 8*, 4593–4596. https://doi.org/10.1021/nl8029114
11. Smith, A., Duan, H., Mohs, A., & Nie, S. (2008). Bioconjugated quantum dots for in vivo molecular and cellular imaging☆. *Advanced Drug Delivery Reviews, 60*, 1226–1240. https://doi.org/10.1016/j.addr.2008.03.015
12. Grandhi, G. K., & Viswanatha, A. M. R. (2016). Understanding the role of surface capping ligands in passivating the quantum dots using copper dopants as internal sensor. *The Journal of Physical Chemistry C, 120*, 19785–19795. https://doi.org/10.1021/acs.jpcc.6b04060
13. Elugoke, S. E., Adekunle, A. S., Fayemi, O. E., Mamba, B. B., Sherif, E.-S. M., & Ebenso, E. E. (2020). Carbon-based quantum dots for electrochemical detection of monoamine neurotransmitters—Review. *Biosensors (Basel), 10*, 162. https://doi.org/10.3390/bios10110162
14. Chen, Y., Jiang, H., Liu, X., & Wang, X. (2023). Engineered Electrochemiluminescence biosensors for monitoring heavy metal ions: Current status and prospects. *Biosensors (Basel), 14*, 9. https://doi.org/10.3390/bios14010009
15. Chern, M., Kays, J. C., Bhuckory, S., & Dennis, A. M. (2019). Sensing with photoluminescent semiconductor quantum dots. *Methods and Applications in Fluorescence, 7*, 012005. https://doi.org/10.1088/2050-6120/aaf6f8
16. Xu, L., Li, J., Cai, B., Song, J., Zhang, F., Fang, T., & Zeng, H. (2020). A bilateral interfacial passivation strategy promoting efficiency and stability of perovskite quantum dot light-emitting diodes. *Nature Communications, 11*, 3902. https://doi.org/10.1038/s41467-020-17633-3
17. Frasco, M. F., & Chaniotakis, N. (2009). Semiconductor quantum dots in chemical sensors and biosensors. *Sensors, 9*, 7266–7286. https://doi.org/10.3390/s90907266
18. Verma, A. K., Noumani, A., Yadav, A. K., & Solanki, P. R. (2023). FRET based biosensor: Principle applications recent advances and challenges. *Diagnostics, 13*, 1375. https://doi.org/10.3390/diagnostics13081375
19. Suzuki, M., Husimi, Y., Komatsu, H., Suzuki, K., & Douglas, K. T. (2008). Quantum dot FRET biosensors that respond to pH, to proteolytic or nucleolytic cleavage, to DNA synthesis, or to a multiplexing combination. *Journal of the American Chemical Society, 130*, 5720–5725. https://doi.org/10.1021/ja710870e
20. Ramanarayanan, R., Ummer, F. C., & Swaminathan, S. (2020). Exploring dynamics of resonance energy transfer in hybrid quantum dot sensitized solar cells (QDSSC). *Materials Research Express, 7*, 025517. https://doi.org/10.1088/2053-1591/ab761b
21. Gu, C. (2017). Quantum dots-based fluorescence resonance energy transfer biosensor for monitoring cell apoptosis. *Luminescence, 32*, 1186–1191. https://doi.org/10.1002/bio.3309
22. Lu, X., Hou, X., Tang, H., Yi, X., & Wang, J. (2022). A high-quality CdSe/CdS/ZnS quantum-dot-based FRET Aptasensor for the simultaneous detection of two different Alzheimer's disease Core biomarkers. *Nanomaterials, 12*, 4031. https://doi.org/10.3390/nano12224031
23. Yu, H., Alkhamis, O., Canoura, J., Liu, Y., & Xiao, Y. (2021). Advances and challenges in small-molecule DNA aptamer isolation, characterization, and sensor development. *Angewandte Chemie, 133*, 16938–16961. https://doi.org/10.1002/ange.202008663
24. Rahmawati, I., Einaga, Y., Ivandini, T. A., & Fiorani, A. (2022). Enzymatic biosensors with electrochemiluminescence transduction. *ChemElectroChem, 9*. https://doi.org/10.1002/celc.202200175
25. Farzin, M. A., & Abdoos, H. (2021). A critical review on quantum dots: From synthesis toward applications in electrochemical biosensors for determination of disease-related biomolecules. *Talanta, 224*, 121828. https://doi.org/10.1016/j.talanta.2020.121828
26. Mokwebo, K. V., Oluwafemi, O. S., & Arotiba, O. A. (2018). An electrochemical cholesterol biosensor based on a CdTe/CdSe/ZnSe quantum dots—Poly (propylene imine) dendrimer nanocomposite immobilization layer. *Sensors, 18*, 3368. https://doi.org/10.3390/s18103368
27. Cai, J., Sun, B., Gou, X., Gou, Y., Li, W., & Hu, F. (2018). A novel way for analysis of calycosin via polyaniline functionalized graphene quantum dots fabricated electrochemi-

cal sensor. *Journal of Electroanalytical Chemistry, 816*, 123–131. https://doi.org/10.1016/j.jelechem.2018.03.035
28. Arumugasamy, S. K., Govindaraju, S., & Yun, K. (2020). Electrochemical sensor for detecting dopamine using graphene quantum dots incorporated with multiwall carbon nanotubes. *Applied Surface Science, 508*, 145294. https://doi.org/10.1016/j.apsusc.2020.145294
29. Wu, X., Ma, P., Sun, Y., Du, F., Song, D., & Xu, G. (2021). Application of MXene in electrochemical sensors: A review. *Electroanalysis, 33*, 1827–1851. https://doi.org/10.1002/elan.202100192
30. Jung, E.-Y., Ye, J.-H., Jung, S.-H., & Choi, S.-H. (2016). Electrochemiluminescence biosensor based on thioglycolic acid-capped CdSe QDs for sensing glucose. *Journal of Nanomaterials, 2016*, 1–9. https://doi.org/10.1155/2016/5760327
31. He, Z.-J., Kang, T.-F., Lu, L.-P., & Cheng, S.-Y. (2020). An electrochemiluminescence aptamer sensor for chloramphenicol based on GO-QDs nanocomposites and enzyme-linked aptamers. *Journal of Electroanalytical Chemistry, 860*, 113870. https://doi.org/10.1016/j.jelechem.2020.113870
32. Mostafa, M., El Nady, J., Ebrahim, S. M., & Elshaer, A. M. (2021). Synthesis, structural, and optical properties of Mn2+ doped ZnS quantum dots for biosensor application. *Optical Materials (Amst), 112*, 110732. https://doi.org/10.1016/j.optmat.2020.110732
33. Safardoust-Hojaghan, H., Salavati-Niasari, M., Amiri, O., & Hassanpour, M. (2017). Preparation of highly luminescent nitrogen doped graphene quantum dots and their application as a probe for detection of *Staphylococcus aureus* and *E. coli*. *Journal of Molecular Liquids, 241*, 1114–1119. https://doi.org/10.1016/j.molliq.2017.06.106
34. Hasanzadeh, M., Tagi, S., Solhi, E., Mokhtarzadeh, A., Shadjou, N., Eftekhari, A., & Mahboob, S. (2018). An innovative immunosensor for ultrasensitive detection of breast cancer specific carbohydrate (CA 15-3) in unprocessed human plasma and MCF-7 breast cancer cell lysates using gold nanospear electrochemically assembled onto thiolated graphene quantum dots. *International Journal of Biological Macromolecules, 114*, 1008–1017. https://doi.org/10.1016/j.ijbiomac.2018.03.183
35. Sun, B., Wang, Y., Li, D., Li, W., Gou, X., Gou, Y., & Hu, F. (2020). Development of a sensitive electrochemical immunosensor using polyaniline functionalized graphene quantum dots for detecting a depression marker. *Materials Science and Engineering: C, 111*, 110797. https://doi.org/10.1016/j.msec.2020.110797
36. Aruguete, D. M., Guest, J. S., Yu, W. W., Love, N. G., & Hochella, M. F. (2010). Interaction of CdSe/CdS core-shell quantum dots and *Pseudomonas aeruginosa*. *Environmental Chemistry, 7*, 28. https://doi.org/10.1071/EN09106
37. Ehzari, H., Amiri, M., & Safari, M. (2020). Enzyme-free sandwich-type electrochemical immunosensor for highly sensitive prostate specific antigen based on conjugation of quantum dots and antibody on surface of modified glassy carbon electrode with core–shell magnetic metal-organic frameworks. *Talanta, 210*, 120641. https://doi.org/10.1016/j.talanta.2019.120641
38. Zhao, Y., Tan, L., & Jie, G. (2021). Ultrasensitive electrochemiluminescence biosensor for the detection of carcinoembryonic antigen based on multiple amplification and a DNA walker. *Sensors and Actuators B: Chemical, 333*, 129586. https://doi.org/10.1016/j.snb.2021.129586
39. Zhang, Q., Liu, Y., Nie, Y., Liu, Y., & Ma, Q. (2019). Wavelength-dependent surface plasmon coupling electrochemiluminescence biosensor based on sulfur-doped carbon nitride quantum dots for K-RAS gene detection. *Analytical Chemistry, 91*, 13780–13786. https://doi.org/10.1021/acs.analchem.9b03212
40. Jin, R., Lin, B., Li, D., & Ai, H. (2014). Superparamagnetic iron oxide nanoparticles for MR imaging and therapy: Design considerations and clinical applications. *Current Opinion in Pharmacology, 18*, 18–27. https://doi.org/10.1016/j.coph.2014.08.002

41. Ahmad, J., Garg, A., Mustafa, G., Ahmad, M. Z., Aslam, M., & Mishra, A. (2023). Hybrid quantum dot as promising tools for Theranostic application in cancer. *Electronics (Basel), 12*, 972. https://doi.org/10.3390/electronics12040972
42. Kashyap, B. K., Singh, V. V., Solanki, M. K., Kumar, A., Ruokolainen, J., & Kesari, K. K. (2023). Smart nanomaterials in cancer theranostics: Challenges and opportunities. *ACS Omega, 8*, 14290–14320. https://doi.org/10.1021/acsomega.2c07840
43. Yeh, C.-Y., Hsiao, J.-K., Wang, Y.-P., Lan, C.-H., & Wu, H.-C. (2016). Peptide-conjugated nanoparticles for targeted imaging and therapy of prostate cancer. *Biomaterials, 99*, 1–15. https://doi.org/10.1016/j.biomaterials.2016.05.015
44. Huang, N., Cheng, S., Zhang, X., Tian, Q., Pi, J., Tang, J., Huang, Q., Wang, F., Chen, J., Xie, Z., Xu, Z., Chen, W., Zheng, H., & Cheng, Y. (2017). Efficacy of NGR peptide-modified PEGylated quantum dots for crossing the blood–brain barrier and targeted fluorescence imaging of glioma and tumor vasculature. *Nanomedicine, 13*, 83–93. https://doi.org/10.1016/j.nano.2016.08.029
45. Balalaeva, I. V., Zdobnova, T. A., Krutova, I. V., Brilkina, A. A., Lebedenko, E. N., & Deyev, S. M. (2012). Passive and active targeting of quantum dots for whole-body fluorescence imaging of breast cancer xenografts. *Journal of Biophotonics, 5*, 860–867. https://doi.org/10.1002/jbio.201200080
46. Li, S., Su, W., Wu, H., Yuan, T., Yuan, C., Liu, J., Deng, G., Gao, X., Chen, Z., Bao, Y., Yuan, F., Zhou, S., Tan, H., Li, Y., Li, X., Fan, L., Zhu, J., Chen, A. T., Liu, F., Zhou, Y., Li, M., Zhai, X., & Zhou, J. (2020). Targeted tumour theranostics in mice via carbon quantum dots structurally mimicking large amino acids. *Nature Biomedical Engineering, 4*, 704–716. https://doi.org/10.1038/s41551-020-0540-y
47. Kamba, S., Ismail, M., Hussein-Al-Ali, S., Ibrahim, T., & Zakaria, Z. (2013). In vitro delivery and controlled release of doxorubicin for targeting osteosarcoma bone cancer. *Molecules, 18*, 10580–10598. https://doi.org/10.3390/molecules180910580
48. Fakayode, O. J., Tsolekile, N., Songca, S. P., & Oluwafemi, O. S. (2018). Applications of functionalized nanomaterials in photodynamic therapy. *Biophysical Reviews, 10*, 49–67. https://doi.org/10.1007/s12551-017-0383-2
49. Hwang, S., Nam, J., Jung, S., Song, J., Doh, H., & Kim, S. (2014). Gold nanoparticle-mediated photothermal therapy: Current status and future perspective. *Nanomedicine, 9*, 2003–2022. https://doi.org/10.2217/nnm.14.147
50. Giustini, A. J., Petryk, A. A., Cassim, S. M., Tate, J. A., Baker, I., & Hoopes, P. J. (2010). Magnetic nanoparticle hyperthermia in cancer treatment. *Nano Life, 01*, 17–32. https://doi.org/10.1142/S1793984410000067
51. Bao, X., Yuan, Y., Chen, J., Zhang, B., Li, D., Zhou, D., Jing, P., Xu, G., Wang, Y., Holá, K., Shen, D., Wu, C., Song, L., Liu, C., Zbořil, R., & Qu, S. (2018). In vivo theranostics with near-infrared-emitting carbon dots—Highly efficient photothermal therapy based on passive targeting after intravenous administration. *Light: Science & Applications, 7*, 91. https://doi.org/10.1038/s41377-018-0090-1
52. Wang, L., Xu, D., Gao, J., Chen, X., Duo, Y., & Zhang, H. (2020). Semiconducting quantum dots: Modification and applications in biomedical science. *Science China Materials, 63*, 1631–1650. https://doi.org/10.1007/s40843-020-1330-7
53. Yao, V. J., D'Angelo, S., Butler, K. S., Theron, C., Smith, T. L., Marchiò, S., Gelovani, J. G., Sidman, R. L., Dobroff, A. S., Brinker, C. J., Bradbury, A. R. M., Arap, W., & Pasqualini, R. (2016). Ligand-targeted theranostic nanomedicines against cancer. *Journal of Controlled Release, 240*, 267–286. https://doi.org/10.1016/j.jconrel.2016.01.002
54. Pan, J., & Feng, S.-S. (2009). Targeting and imaging cancer cells by folate-decorated, quantum dots (QDs)-loaded nanoparticles of biodegradable polymers. *Biomaterials, 30*, 1176–1183. https://doi.org/10.1016/j.biomaterials.2008.10.039
55. Alibolandi, M., Abnous, K., Sadeghi, F., Hosseinkhani, H., Ramezani, M., & Hadizadeh, F. (2016). Folate receptor-targeted multimodal polymersomes for delivery of quantum dots and doxorubicin to breast adenocarcinoma: In vitro and in vivo evaluation. *International Journal of Pharmaceutics, 500*, 162–178. https://doi.org/10.1016/j.ijpharm.2016.01.040

56. van der Meel, R., Vehmeijer, L. J. C., Kok, R. J., Storm, G., & van Gaal, E. V. B. (2013). Ligand-targeted particulate nanomedicines undergoing clinical evaluation: Current status. *Advanced Drug Delivery Reviews, 65*, 1284–1298. https://doi.org/10.1016/j.addr.2013.08.012
57. Nakamura, Y., Mochida, A., Choyke, P. L., & Kobayashi, H. (2016). Nanodrug delivery: Is the enhanced permeability and retention effect sufficient for curing cancer? *Bioconjugate Chemistry, 27*, 2225–2238. https://doi.org/10.1021/acs.bioconjchem.6b00437
58. Bentolila, L. A., Ebenstein, Y., & Weiss, S. (2009). Quantum dots for in vivo small-animal imaging. *Journal of Nuclear Medicine, 50*, 493–496. https://doi.org/10.2967/jnumed.108.053561
59. Salunkhe, S., Dheeraj, M., Basak, D., & Chitkara, A. M. (2020). Surface functionalization of exosomes for target-specific delivery and in vivo imaging & tracking: Strategies and significance. *Journal of Controlled Release, 326*, 599–614. https://doi.org/10.1016/j.jconrel.2020.07.042
60. Zwicke, G. L., Ali Mansoori, G., & Jeffery, C. J. (2012). Utilizing the folate receptor for active targeting of cancer nanotherapeutics. *Nano Reviews, 3*, 18496. https://doi.org/10.3402/nano.v3i0.18496
61. He, Z.-Y., Jin, Z.-H., Zhan, M., Qin, Z., Li, Z., & Xu, T. (2017). Advances in quantum dot-mediated siRNA delivery. *Chinese Chemical Letters, 28*, 1851–1856. https://doi.org/10.1016/j.cclet.2017.07.012
62. Xu, B., Cao, Q., Zhang, Y., Yu, W., Zhu, J., Liu, D., & Jiang, G. (2018). Microneedles integrated with ZnO quantum-dot-capped mesoporous bioactive glasses for glucose-mediated insulin delivery. *ACS Biomaterials Science & Engineering, 4*, 2473–2483. https://doi.org/10.1021/acsbiomaterials.8b00626
63. Attia, M. F., Anton, N., Wallyn, J., Omran, Z., & Vandamme, T. F. (2019). An overview of active and passive targeting strategies to improve the nanocarriers efficiency to tumour sites. *Journal of Pharmacy and Pharmacology, 71*, 1185–1198. https://doi.org/10.1111/jphp.13098
64. Kar, R., Dhar, R., Mukherjee, S., Nag, S., Gorai, S., Mukerjee, N., Mukherjee, D., Vatsa, R., Chandrakanth Jadhav, M., Ghosh, A., Devi, A., Krishnan, A., & Thorat, N. D. (2023). Exosome-based smart drug delivery tool for cancer theranostics. *ACS Biomaterials Science & Engineering, 9*, 577–594. https://doi.org/10.1021/acsbiomaterials.2c01329
65. Guo, S., & Huang, L. (2011). Nanoparticles escaping RES and endosome: Challenges for siRNA delivery for cancer therapy. *Journal of Nanomaterials, 2011*, 1–12. https://doi.org/10.1155/2011/742895
66. Jung, J., Solanki, A., Memoli, K. A., Kamei, K., Kim, H., Drahl, M. A., Williams, L. J., Tseng, H., & Lee, K. (2010). Selective inhibition of human brain tumor cells through multi-functional quantum-dot-based siRNA delivery. *Angewandte Chemie, 122*, 107–111. https://doi.org/10.1002/ange.200905126
67. Lu, X., Zhang, Z., Xia, Q., Hou, M., Yan, C., Chen, Z., Xu, Y., & Liu, R. (2018). Glucose functionalized carbon quantum dot containing organic radical for optical/MR dual-modality bioimaging. *Materials Science and Engineering: C, 82*, 190–196. https://doi.org/10.1016/j.msec.2017.08.074
68. Derfus, A. M., Chan, W. C. W., & Bhatia, S. N. (2004). Probing the cytotoxicity of semiconductor quantum dots. *Nano Letters, 4*, 11–18. https://doi.org/10.1021/nl0347334
69. Paesano, L., Perotti, A., Buschini, A., Carubbi, C., Marmiroli, M., Maestri, E., Iannotta, S., & Marmiroli, N. (2016). Markers for toxicity to HepG2 exposed to cadmium sulfide quantum dots; damage to mitochondria. *Toxicology, 374*, 18–28. https://doi.org/10.1016/j.tox.2016.11.012
70. Yu, K.-N., Sung, J. H., Lee, S., Kim, J.-E., Kim, S., Cho, W.-Y., Lee, A. Y., Park, S. J., Lim, J., Park, C., Chae, C., Lee, J. K., Lee, J., Kim, J.-S., & Cho, M.-H. (2015). Inhalation of titanium dioxide induces endoplasmic reticulum stress-mediated autophagy and inflammation in mice. *Food and Chemical Toxicology, 85*, 106–113. https://doi.org/10.1016/j.fct.2015.08.001
71. Kumar, S., Meena, R., & Paulraj, R. (2016). Role of macrophage (M1 and M2) in titanium-dioxide nanoparticle-induced oxidative stress and inflammatory response in

rat. *Applied Biochemistry and Biotechnology, 180*, 1257–1275. https://doi.org/10.1007/s12010-016-2165-x

72. Yong, K.-T., & Swihart, M. T. (2012). In vivo toxicity of quantum dots: No cause for concern? *Nanomedicine, 7*, 1641–1643. https://doi.org/10.2217/nnm.12.152
73. Lu, J., Tang, M., & Zhang, T. (2019). Review of toxicological effect of quantum dots on the liver. *Journal of Applied Toxicology, 39*, 72–86. https://doi.org/10.1002/jat.3660
74. Yang, Y., Li, P., Lin, Y., Li, Z., Cui, T., Song, Z., Wu, W., Lv, S., & Ji, S. (2020). <p>gene expression profiling of the liver and lung in mice after exposure to ZnO quantum dots</p>. *International Journal of Nanomedicine, 15*, 2947–2955. https://doi.org/10.2147/IJN.S246754
75. Zhao, L., Zong, W., Zhang, H., & Liu, R. (2019). Kidney toxicity and response of selenium containing protein-glutathione peroxidase (Gpx3) to CdTe QDs on different levels. *Toxicological Sciences, 168*, 201–208. https://doi.org/10.1093/toxsci/kfy297
76. Huang, P., Wang, M., Wang, J., Sun, H., Han, S., Feng, S., Shi, L., Meng, P., Li, J., & Sun, Z. (2016). Time-dependent toxicity of cadmium telluride quantum dots on liver and kidneys in mice: Histopathological changes with elevated free cadmium ions and hydroxyl radicals. *International Journal of Nanomedicine, 2016*, 2319. https://doi.org/10.2147/IJN.S103489
77. Ma-Hock, L., Farias, P. M. A., Hofmann, T., Andrade, A. C. D. S., Silva, J. N., Arnaud, T. M. S., Wohlleben, W., Strauss, V., Treumann, S., Chaves, C. R., Gröters, S., Landsiedel, R., & van Ravenzwaay, B. (2014). Short term inhalation toxicity of a liquid aerosol of glutaraldehyde-coated CdS/Cd(OH)2 core shell quantum dots in rats. *Toxicology Letters, 225*, 20–26. https://doi.org/10.1016/j.toxlet.2013.11.007
78. Jacobsen, N. R., Møller, P., Jensen, K. A., Vogel, U., Ladefoged, O., Loft, S., & Wallin, H. (2009). Lung inflammation and genotoxicity following pulmonary exposure to nanoparticles in ApoE−/− mice. *Particle and Fibre Toxicology, 6*, 2. https://doi.org/10.1186/1743-8977-6-2
79. Wu, T., He, K., Zhan, Q., Ang, S., Ying, J., Zhang, S., Zhang, T., Xue, Y., Chen, Y., & Tang, M. (2015). Partial protection of N-acetylcysteine against MPA-capped CdTe quantum dot-induced neurotoxicity in rat primary cultured hippocampal neurons. *Toxicology Research (Camb), 4*, 1613–1622. https://doi.org/10.1039/C5TX00127G
80. Hopkins, L. E., Patchin, E. S., Chiu, P.-L., Brandenberger, C., Smiley-Jewell, S., & Pinkerton, K. E. (2014). Nose-to-brain transport of aerosolized quantum dots following acute exposure. *Nanotoxicology, 8*, 885–893. https://doi.org/10.3109/17435390.2013.842267
81. Zhang, M., Bishop, B. P., Thompson, N. L., Hildahl, K., Dang, B., Mironchuk, O., Chen, N., Aoki, R., Holmberg, V. C., & Nance, E. (2019). Quantum dot cellular uptake and toxicity in the developing brain: Implications for use as imaging probes. *Nanoscale Advances, 1*, 3424–3442. https://doi.org/10.1039/C9NA00334G
82. Misra, R., Das, M., Biswas, P., & Nanda, A. (2021). EGFR targeted Mn-doped ZnO fluorescent nanocrystals for cancer theranostic application. *Materials Today, Communications, 26*, 102170. https://doi.org/10.1016/j.mtcomm.2021.102170
83. Mancini, M. C., Kairdolf, B. A., Smith, A. M., & Nie, S. (2008). Oxidative quenching and degradation of polymer-encapsulated quantum dots: new insights into the long-term fate and toxicity of nanocrystals in vivo. *Journal of the American Chemical Society, 130*, 10836–10837. https://doi.org/10.1021/ja8040477
84. Anilkumar, M., Bindu, K. R., Sneha Saj, A., & Anila, E. I. (2016). Enhanced biocompatibility of ZnS:Mn quantum dots encapsulated with *Aloe vera* extract for therapeutic applications. *Chinese Physics B, 25*, 088103. https://doi.org/10.1088/1674-1056/25/8/088103
85. Nasrollahi, F., Sana, B., Paramelle, D., Ahadian, S., Khademhosseini, A., & Lim, S. (2020). Incorporation of graphene quantum dots, iron, and doxorubicin in/on ferritin nanocages for bimodal imaging and drug delivery. *Advanced Therapeutics (Weinh), 3*. https://doi.org/10.1002/adtp.201900183
86. Shpyleva, S. I., Tryndyak, V. P., Kovalchuk, O., Starlard-Davenport, A., Chekhun, V. F., Beland, F. A., & Pogribny, I. P. (2011). Role of ferritin alterations in human breast cancer cells. *Breast Cancer Research and Treatment, 126*, 63–71. https://doi.org/10.1007/s10549-010-0849-4

87. Yu Lee, J., Heon Nam, D., Hwa Oh, M., Kim, Y., Seok Choi, H., Young Jeon, D., Beum Park, C., & Sung Nam, Y. (2014). Serum-stable quantum dot—Protein hybrid nanocapsules for optical bioimaging. *Nanotechnology, 25*, 175702. https://doi.org/10.1088/0957-4484/25/17/175702
88. Tan, W. B., Huang, N., & Zhang, Y. (2007). Ultrafine biocompatible chitosan nanoparticles encapsulating multicoloured quantum dots for bioapplications. *Journal of Colloid and Interface Science, 310*, 464–470. https://doi.org/10.1016/j.jcis.2007.01.083
89. Chen, L., Willoughby, A., & Zhang, J. (2014). Luminescent gelatin nanospheres by encapsulating CdSe quantum dots. *Luminescence, 29*, 74–78. https://doi.org/10.1002/bio.2505
90. Sheng, W., Kim, S., Lee, J., Kim, S.-W., Jensen, K., & Bawendi, M. G. (2006). In-situ encapsulation of quantum dots into polymer microspheres. *Langmuir, 22*, 3782–3790. https://doi.org/10.1021/la051973l
91. Romoser, A., Ritter, D., Majitha, R., Meissner, K. E., McShane, M., & Sayes, C. M. (2011). Mitigation of quantum dot cytotoxicity by microencapsulation. *PLoS One, 6*, e22079. https://doi.org/10.1371/journal.pone.0022079
92. Bruun, K., & Hille, C. (2019). Study on intracellular delivery of liposome encapsulated quantum dots using advanced fluorescence microscopy. *Scientific Reports, 9*, 10504. https://doi.org/10.1038/s41598-019-46732-5
93. Dudu, V., Ramcharan, M., Gilchrist, M. L., Holland, E. C., & Vazquez, M. (2008). Liposome delivery of quantum dots to the cytosol of live cells. *Journal of Nanoscience and Nanotechnology, 8*, 2293–2300. https://doi.org/10.1166/jnn.2008.185
94. Galloway, J. F., Winter, A., Lee, K. H., Park, J. H., Dvoracek, C. M., Devreotes, P., & Searson, P. C. (2012). Quantitative characterization of the lipid encapsulation of quantum dots for biomedical applications. *Nanomedicine, 8*, 1190–1199. https://doi.org/10.1016/j.nano.2011.12.002
95. Juzenas, P., Generalov, R., Kavaliauskiene, S., Kristensen, S., Chen, & Westrøm, M. (2011). Entrapment in phospholipid vesicles quenches photoactivity of quantum dots. *International Journal of Nanomedicine, 6*, 1875. https://doi.org/10.2147/IJN.S22953
96. Souris, J. S., Lee, C.-H., Cheng, S.-H., Chen, C.-T., Yang, C.-S., Ho, J. A., Mou, C.-Y., & Lo, L.-W. (2010). Surface charge-mediated rapid hepatobiliary excretion of mesoporous silica nanoparticles. *Biomaterials, 31*, 5564–5574. https://doi.org/10.1016/j.biomaterials.2010.03.048
97. Hu, X., Zrazhevskiy, P., & Gao, X. (2009). Encapsulation of single quantum dots with mesoporous silica. *Annals of Biomedical Engineering, 37*, 1960–1966. https://doi.org/10.1007/s10439-009-9660-y
98. Asefa, T., & Tao, Z. (2012). Biocompatibility of mesoporous silica nanoparticles. *Chemical Research in Toxicology, 25*, 2265–2284. https://doi.org/10.1021/tx300166u
99. Huang, X., Teng, X., Chen, D., Tang, F., & He, J. (2010). The effect of the shape of mesoporous silica nanoparticles on cellular uptake and cell function. *Biomaterials, 31*, 438–448. https://doi.org/10.1016/j.biomaterials.2009.09.060
100. Kim, Y.-H., Subramanyam, E., Im, J. H., Huh, K. M., Choi, H., Choi, J. S., Lee, Y.-K., & Park, S.-W. (2010). A new PEG-lipid conjugate micelle for encapsulation of CdSe/ZnS quantum dots. *Journal of Nanoscience and Nanotechnology, 10*, 3275–3279. https://doi.org/10.1166/jnn.2010.2230
101. Hu, X., & Gao, X. (2010). Silica–polymer dual layer-encapsulated quantum dots with remarkable stability. *ACS Nano, 4*, 6080–6086. https://doi.org/10.1021/nn1017044
102. Tomczak, N., Liu, R., & Vancso, J. G. (2013). Polymer-coated quantum dots. *Nanoscale, 5*, 12018. https://doi.org/10.1039/c3nr03949h
103. Smith, W. E., Brownell, J., White, C. C., Afsharinejad, Z., Tsai, J., Hu, X., Polyak, S. J., Gao, X., Kavanagh, T. J., & Eaton, D. L. (2012). In vitro toxicity assessment of Amphiphillic polymer-coated CdSe/ZnS quantum dots in two human liver cell models. *ACS Nano, 6*, 9475–9484. https://doi.org/10.1021/nn302288r

104. Wang, W., Kapur, A., Ji, X., Zeng, B., Mishra, D., & Mattoussi, H. (2016). Multifunctional and high affinity polymer ligand that provides bio-orthogonal coating of quantum dots. *Bioconjugate Chemistry, 27*, 2024–2036. https://doi.org/10.1021/acs.bioconjchem.6b00309
105. Wang, W., Ji, X., Kapur, A., Zhang, C., & Mattoussi, H. (2015). A multifunctional polymer combining the imidazole and zwitterion motifs as a biocompatible compact coating for quantum dots. *Journal of the American Chemical Society, 137*, 14158–14172. https://doi.org/10.1021/jacs.5b08915
106. Susumu, K., Oh, E., Delehanty, J. B., Blanco-Canosa, J. B., Johnson, B. J., Jain, V., Hervey, W. J., Algar, W. R., Boeneman, K., Dawson, P. E., & Medintz, I. L. (2011). Multifunctional compact zwitterionic ligands for preparing robust biocompatible semiconductor quantum dots and gold nanoparticles. *Journal of the American Chemical Society, 133*, 9480–9496. https://doi.org/10.1021/ja201919s
107. Sun, M., Yang, L., Jose, P., Wang, L., & Zweit, J. (2013). Functionalization of quantum dots with multidentate zwitterionic ligands: Impact on cellular interactions and cytotoxicity. *Journal of Materials Chemistry B, 1*, 6137. https://doi.org/10.1039/c3tb20894j
108. Wei, Y., Jana, N. R., Tan, S. J., & Ying, J. Y. (2009). Surface coating directed cellular delivery of TAT-functionalized quantum dots. *Bioconjugate Chemistry, 20*, 1752–1758. https://doi.org/10.1021/bc8003777
109. Xu, J., Ruchala, P., Ebenstain, Y., Li, J. J., & Weiss, S. (2012). Stable, compact, bright biofunctional quantum dots with improved peptide coating. *The Journal of Physical Chemistry. B, 116*, 11370–11378. https://doi.org/10.1021/jp306453y
110. Selvan, S. T., Tan, T. T., & Ying, J. Y. (2005). Robust, non-cytotoxic, silica-coated CdSe quantum dots with efficient photoluminescence. *Advanced Materials, 17*, 1620–1625. https://doi.org/10.1002/adma.200401960
111. Ma, Y., Li, Y., Ma, S., & Zhong, X. (2014). Highly bright water-soluble silica coated quantum dots with excellent stability. *Journal of Materials Chemistry B, 2*, 5043–5051. https://doi.org/10.1039/C4TB00458B
112. Liu, Y., Ai, K., Yuan, Q., & Lu, L. (2011). Fluorescence-enhanced gadolinium-doped zinc oxide quantum dots for magnetic resonance and fluorescence imaging. *Biomaterials, 32*, 1185–1192. https://doi.org/10.1016/j.biomaterials.2010.10.022
113. Ramalho, J., Ramalho, M., Jay, M., Burke, L. M., & Semelka, R. C. (2016). Gadolinium toxicity and treatment. *Magnetic Resonance Imaging, 34*, 1394–1398. https://doi.org/10.1016/j.mri.2016.09.005
114. Marković, Z. M., Labudová, M., Danko, M., Matijašević, D., Mičušík, M., Nádaždy, V., Kováčová, M., Kleinová, A., Špitalský, Z., Pavlović, V., Milivojević, D. D., Medić, M., & Todorović Marković, B. M. (2020). Highly efficient antioxidant F- and cl-doped carbon quantum dots for bioimaging. *ACS Sustainable Chemistry & Engineering, 8*, 16327–16338. https://doi.org/10.1021/acssuschemeng.0c06260
115. Kyzer, J. L., & Martens, M. (2021). Metabolism and toxicity of fluorine compounds. *Chemical Research in Toxicology, 34*, 678–680. https://doi.org/10.1021/acs.chemrestox.0c00439
116. Winder, C. (2001). The toxicology of chlorine. *Environmental Research, 85*, 105–114. https://doi.org/10.1006/enrs.2000.4110
117. Qian, Z., Shan, X., Chai, L., Ma, J., Chen, J., & Feng, H. (2014). Si-doped carbon quantum dots: A facile and general preparation strategy, bioimaging application, and multifunctional sensor. *ACS Applied Materials & Interfaces, 6*, 6797–6805. https://doi.org/10.1021/am500403n
118. Kumar, V. B., Kumar, R., Gedanken, A., & Shefi, O. (2019). Fluorescent metal-doped carbon dots for neuronal manipulations. *Ultrasonics Sonochemistry, 52*, 205–213. https://doi.org/10.1016/j.ultsonch.2018.11.017
119. Zhu, C., Chen, Z., Gao, S., Goh, B. L., Bin Samsudin, I., Lwe, K. W., Wu, Y., Wu, C., & Su, X. (2019). Recent advances in nontoxic quantum dots and their biomedical applications. *Progress in Natural Science: Materials International, 29*, 628–640. https://doi.org/10.1016/j.pnsc.2019.11.007

120. Chinnathambi, S., Shirahata, N., Lesani, P., Thangavel, V., & Pandian, G. N. (2024). Surface charge-dependent cytokine production using near-infrared emitting silicon quantum dots. *Scientific Reports, 14*, 9618. https://doi.org/10.1038/s41598-024-60536-2
121. Tang, Y., Han, S., Liu, H., Chen, X., Huang, L., Li, X., & Zhang, J. (2013). The role of surface chemistry in determining in vivo biodistribution and toxicity of CdSe/ZnS core–shell quantum dots. *Biomaterials, 34*, 8741–8755. https://doi.org/10.1016/j.biomaterials.2013.07.087
122. Lovrić, J., Bazzi, H. S., Cuie, Y., Fortin, G. R. A., Winnik, F. M., & Maysinger, D. (2005). Differences in subcellular distribution and toxicity of green and red emitting CdTe quantum dots. *Journal of Molecular Medicine, 83*, 377–385. https://doi.org/10.1007/s00109-004-0629-x

Quantum Dot Localized Surface Plasmon Resonance Biomedical Applications

E. Vignesh Balaji and K. Sreedhara Ranganath Pai

1 Introduction

Quantum dots (QDs) are semiconductor nanocrystals with remarkable optical and electronic properties arising from their size and shape at the nanoscale. Typically, quantum dots range from 1 to 10 nm in diameter, which is small enough to cause quantum confinement effects. These effects occur when the motion of electrons or holes is restricted to one or more spatial dimensions, leading to discrete energy levels, similar to those of an atom. The quantum confinement of quantum dots allows them to have unique electronic and optical behaviors, making them a subject of intense research across various scientific fields, particularly in physics, chemistry, materials science, and engineering.

At the core of the distinctive characteristics of quantum dots are their size-dependent optical properties. When excited by light, QDs can absorb photons, and depending on their size, they emit light of different wavelengths. Smaller quantum dots typically emit light at shorter wavelengths (blue light), whereas larger dots emit light at longer wavelengths (red light). This size-tunable fluorescence makes quantum dots incredibly valuable for applications in biological imaging, optoelectronics, and photovoltaics, among others. Their emission spectra can be precisely tuned by controlling their size during synthesis, offering a high level of customization for various uses.

The synthesis of quantum dots is a complex process that involves the creation of nanoparticles from semiconductor materials such as cadmium telluride (CdTe), cadmium selenide (CdSe) or indium phosphide (InP), among others. Chemical methods

E. Vignesh Balaji · K. S. R. Pai (✉)
Department of Pharmacology, Manipal College of Pharmaceutical Sciences, Manipal Academy of Higher Education, Manipal, Karnataka, India
e-mail: ksr.pai@manipal.edu

S. P. Kabekkodu et al. (eds.), *Functional Quantum Dots*, Lecture Notes in Nanoscale Science and Technology 40,
https://doi.org/10.1007/978-3-031-98219-4_5

such as colloidal synthesis are commonly employed, where a solution of precursor material is heated to induce the formation of QDs. The size and surface characteristics of quantum dots can be modified by adjusting response conditions such as temperature, solvent, and precursor concentrations. Additionally, surface passivation techniques are used to protect QDs from degradation and improve their stability, photoluminescent properties, and quantum efficiency.

One of the most prominent applications of quantum dots is in the field of biology, particularly for imaging and diagnostics. Quantum dots can function with specific biomolecules, enabling them to target and bind to cells or tissues of interest. Their unique optical properties allow for high-resolution imaging and precise tracking in live cells, offering a significant advantage over traditional organic dyes and fluorescent markers. Furthermore, quantum dots are also used in the development of quantum dot-based solar cells, where their ability to absorb light efficiently and convert it into electrical energy is harnessed to improve solar energy harvesting.

In optoelectronics, quantum dots are used in the creation of light-emitting diodes (LEDs) and display technologies. Compared with conventional liquid crystal displays (LCDs), quantum dot-based displays offer advantages such as improved color purity and energy efficiency. QDs are also being researched for use in lasers, photodetectors, and other devices that rely on precise light emission and detection.

Despite their significant potential, the widespread use of quantum dots faces challenges. These include concerns regarding the toxicity of materials such as cadmium, which is often used in the synthesis of quantum dots, as well as issues related to their long-term stability and scalability for commercial applications. Ongoing research is focused on developing safer, more environmentally friendly quantum dot materials and improving the reproducibility and efficiency of their synthesis [1].

2 Localized Surface Plasmon Resonance

Localized surface plasmon resonance (LSPR) refers to the collective oscillation of free electrons in noble metal nanoparticles when they interact with incident light. This phenomenon is confined to the nanoparticle surface, leading to significant enhancements in local electromagnetic fields and resulting in sharp absorption and scattering peaks in the nanoparticle extinction spectrum [2]. When nanoparticles with dimensions smaller than the incident light wavelength are irradiated, the oscillating electric field of the light induces coherent oscillations of the surface conduction electrons. The frequency at which these electrons resonate is influenced by factors such as the size, shape, material, and dielectric properties of the surrounding medium of the nanoparticle. This resonant frequency typically falls within the visible spectrum for noble metals such as silver and gold, which contributes to the vibrant colors observed in colloidal nanoparticle solutions.

Over the past decade, developments in the fabrication and functionalization of noble metal nanostructures have significantly propelled the development of LSPR-based applications. Researchers have achieved precise control over nanoparticle morphology, enabling tunable LSPR properties. For example, studies have demonstrated that modifying the sharpness of triangular silver nanoparticles alters their absorption maxima, resulting in color changes from blue to red. The sensitivity of LSPR to adjustments in the local dielectric environment has been harnessed in various sensing applications. The resonance frequency shifts in response to molecular adsorption or refractive index changes near the nanoparticle surface, making LSPR a powerful tool for detecting biomolecular interactions, environmental pollutants, and chemical analytes. This sensitivity has led to the development of LSPR-based biosensors capable of real-time, label-free detection with high spatial resolution [3].

In the realm of biomedicine, LSPR-active nanoparticles are being explored for applications such as targeted delivery, photothermal therapy, and bioimaging. The enhanced near-field effects associated with LSPR can increase the efficacy of photothermal treatments by concentrating light energy at the nanoparticle surface, leading to localized heating and selective destruction of diseased cells.

Recent studies have also focused on integrating LSPR with other nanomaterials to further enhance its properties. For example, research on Cu_2O_{1-x} superlattices with oxygen vacancies has shown that introducing such vacancies can significantly enhance LSPR effects, offering new insights into vacancy doping in semiconductors and LSPR induction in metals oxide nanoparticles [2].

3 Biomedical Applications

3.1 Bioimaging

Quantum dots (QDs) are semiconductor nanoparticles with unique optical properties, such as size-dependent fluorescence, that have made them highly valuable in bioimaging applications. Their ability to emit bright, stable, and tunable fluorescence upon excitation allows for enhanced imaging capabilities in biological systems.

Quantum dots are used as fluorescent probes in live-cell imaging, allowing researchers to track cellular processes with high resolution. Their fluorescence can be tuned by adjusting their size, and they exhibit high quantum yield and stability, making them ideal for long-term imaging studies. Since quantum dots can be proposed to emit at different wavelengths, they can be used for multicolor imaging. Multiple quantum dots can be used instantaneously to label different targets within a biological sample, enabling complex, multitarget visualization in a single experiment. Quantum dots can be functionalized with specific biomolecules (such as peptides or antibodies), allowing them to target particular cells, tissues, or even specific proteins. This makes them useful for molecular diagnostics and the study of disease mechanisms. The photostability and deep tissue penetration of quantum dots make them suitable for in vivo imaging. Their small size enables them to navigate through biological systems, allowing noninvasive imaging in living organisms. Quantum dots have been explored in cancer imaging, where they can be conjugated with targeting ligands to specifically bind to tumor cells, providing high-resolution images for diagnosis and assisting in monitoring tumor growth or response to treatment [4].

3.2 Biosensing

QDs are broadly used as fluorescent probes in biological imaging and diagnostics. Their narrow emission spectra and size-dependent properties enable the detection of specific biomolecules in complex biological environments. They are particularly useful in multiplexed assays, where multiple QDs with distinct emission wavelengths can be used simultaneously to detect various targets (e.g., proteins, nucleic acids) in a single sample [5]. Quantum dots are incorporated into biosensors for pathogen detection, especially in immunoassays and DNA hybridization tests. For example, QDs can be functionalized with nucleic acid or antibody probes to specifically recognize pathogens such as bacteria and viruses. This enables rapid, sensitive, and cost-effective pathogen detection, which is crucial for clinical diagnostics and environmental monitoring [6]. QDs are used for real-time cellular biosensing, monitoring cell behavior, and studying molecular interactions within living cells. By tagging biomolecules of interest, QDs can track cellular processes, such as receptor–ligand interactions, protein localization, and gene expression [6]. Owing to their minute size, stability, and ease of functionalization, QDs are being explored for point-of-care (POC) biosensors. These sensors enable rapid, onsite detection of biomarkers without the need for extensive laboratory equipment, making them ideal for low-resource settings and emergency situations [6].

3.3 Drug Delivery

Quantum dots can be functionalized with specific antibodies or ligands that recognize and bind to overexpressed receptors on cancerous or diseased cells. This facilitates targeted drug delivery, reducing the side effects associated with traditional therapies. The surface of QDs can be modified to attach targeting moieties, such as peptides, antibodies, or folic acid, which direct the QDs to specific cells, tissues, or organs. QDs can be used as carriers for controlled drug release. Drugs can be loaded into the quantum dot structure or attached to the surface through various interactions, such as electrostatic forces or covalent bonding. Drug release can be controlled by external stimuli, such as changes in pH, temperature, or light, ensuring that the drug is delivered at the site of action in a controlled manner. Quantum dots are highly fluorescent and can be used for imaging purposes, allowing real-time tracking of drug delivery processes. Their fluorescence properties enable precise monitoring of the drug's location and distribution in the body, aiding in assessing the effectiveness of the delivery system and optimizing the treatment process. This real-time monitoring is particularly useful in personalized medicine. While quantum dots hold immense potential in drug delivery, their toxicity is a concern. Recent research has focused on improving their biocompatibility by using biocompatible coatings such as silica or polymers to reduce harmful interactions with biological systems. Advances in surface functionalization have also made quantum dots safer for in vivo applications [7, 8].

3.4 Biosensing and Diagnostic Applications

LSPR-based biosensors are utilized for detecting biomolecular interactions. The resonance frequency of LSPR shifts in response to changes in the refractive index near the nanoparticle surface, which can occur due to the binding of target molecules (e.g., proteins, nucleic acids, or pathogens). These sensors are highly sensitive, label-free, and offer real-time detection. LSPR sensors have been used to detect biomarkers for cancer with high sensitivity. Recent work has shown the development of LSPR-based sensors capable of sensing low concentrations of tumor-associated biomarkers in blood samples [9]. LSPR can detect viruses and bacteria in clinical samples. For example, a study demonstrated the use of LSPR to rapidly identify pathogens such as *Escherichia coli* and *Staphylococcus aureus*, which are relevant for infection diagnosis [10]. LSPR technology is increasingly used in diagnostic applications because of its high sensitivity, fast detection, and ability to work in complex biological environments. LSPR-based devices are used in portable platforms for the rapid and onsite detection of diseases such as infectious diseases, cardiovascular conditions, and cancer [11].

3.5 Real-Time Detection

LSPR sensors are widely used for detecting biomolecular interactions, such as DNA hybridization, protein binding, or the presence of pathogens, in real time. They offer high sensitivity, allowing for label-free detection with no need for secondary antibodies or fluorescent markers. For example, a real-time LSPR-based biosensor for detecting the SARS-CoV-2 virus provides rapid diagnostics in clinical settings [12]. LSPR sensors can detect pollutants or toxins in real time by measuring the shift in the plasmon resonance caused by target analytes. The use of LSPR for detecting heavy metal ions in water with high specificity and sensitivity has been demonstrated [13]. LSPR has also been applied to detect contaminants, such as pathogens or adulterants, in food products. A study revealed a real-time LSPR sensor capable of detecting foodborne pathogens in agricultural produce [14].

3.6 Cancer Diagnosis

LSPR can be used to detect cancer-specific biomarkers (e.g., proteins, nucleic acids, or glycoproteins) at very low concentrations. The presence of these biomarkers on the surface of nanoparticles causes a shift in plasmon resonance, which can be quantified and linked with the presence of cancer. LSPR-based sensors can be developed into portable, rapid diagnostic devices. These sensors could offer quick, onsite testing for cancer-related biomarkers, reducing the time and cost associated with traditional diagnostic methods. LSPR can be used for real-time examination of cancer progression by detecting specific molecules related to tumor growth and metastasis. Changes in plasmon resonance can indicate changes in the molecular environment as a result of tumor development [15]. Exosomes are small vesicles secreted by tumors into the bloodstream that contain molecular information about the tumor. LSPR can be used to detect exosomes, helping to identify the presence of cancer cells and their genetic information for diagnosis and monitoring [16].

4 Combining Quantum Dots and LSPR in Biomedical Applications

The combination of QDs and LSPR holds great promise for biomedical applications, particularly in imaging, diagnostics, and therapy. Quantum dots are semiconductor nanocrystals that exhibit unique optical properties, including size-tunable fluorescence, due to quantum confinement. LSPR, on the other hand, refers to the resonant oscillation of conduction electrons in nanoparticles, typically metals such as gold or silver, when exposed to light.

In biomedical applications, quantum dots can be conjugated to biomolecules, enabling high-resolution imaging and multiplexed detection because of their stable, bright, and tunable fluorescence. By integrating quantum dots with plasmonic nanoparticles (such as gold or silver), LSPR can enhance the fluorescence of quantum dots, leading to stronger signals, improved sensitivity, and more accurate diagnostics. This combination can also be used in sensing applications, where the shift in the LSPR peak upon biomolecular binding can provide real-time detection of molecular interactions. Additionally, this combination can enhance therapeutic applications. For example, QDs can be used for targeted drug delivery, while LSPR-enhanced photothermal effects from gold nanoparticles can aid in cancer therapy by inducing localized heating and cell destruction [17].

5 Challenges

Many QDs, especially those based on cadmium and lead, exhibit toxicity, raising concerns about their biological and environmental safety. Controlling the shape, size, and surface properties of QDs for consistent performance remains challenging, especially when scaling up production. QDs can undergo photobleaching, meaning that their luminescence diminishes over time when exposed to light, limiting their applications in long-term monitoring and imaging. The LSPR plasmon resonance frequency is highly dependent on the size, shape, and material of the nanoparticle, which can complicate the design of devices with consistent and predictable behavior. LSPR sensors can be sensitive to environmental factors such as temperature and ionic strength, which can affect their detection capabilities in complex biological or chemical media. The precise fabrication of plasmonic nanoparticles and their integration into practical devices, especially for real-world applications such as sensors or imaging, remains challenging.

6 Future Perspectives

Research into less toxic, environmentally friendly QDs (e.g., those based on materials such as carbon, silicon, or nontoxic cadmium-free semiconductors) is ongoing. QDs are being explored for integration into technologies such as quantum computing, photonic devices, and energy harvesting, particularly in tandem with other nanomaterials and technologies. The use of QDs in medical imaging, drug delivery, and biosensing is an exciting area of research, with further improvements in surface functionalization to ensure biocompatibility and stability. Future research may focus on improving the sensitivity of LSPR sensors for applications in environmental monitoring, disease diagnosis, and even food safety testing. The integration of LSPR-based sensors with other materials and techniques, such as microfluidics or optical fibers, could lead to multifunctional devices with enhanced performance. By combining LSPR with machine learning algorithms or adaptive materials, we can create self-calibrating systems capable of detecting a wide range of analytes in real time [18, 19].

7 Conclusion

QDs and LSPR have emerged as powerful tools in biomedical applications owing to their unique optical properties and versatility. Owing to their size-dependent fluorescence, quantum dots allow for high-resolution imaging, enabling precise detection of biological targets at the molecular level. Their surface can be easily modified for functionalization, providing a wide range of possibilities for targeting specific biomolecules, cells, or tissues. On the other hand, LSPR, which is a phenomenon arising from the collective oscillation of conduction electrons on metal nanoparticle surfaces, has proven to be an effective tool for biosensing and diagnostics. The sensitivity of LSPR to changes in the local environment around nanoparticles makes it highly useful for detecting biomolecular interactions, such as antigen–antibody binding or DNA hybridization, without the need for labelling. When combined, quantum dots and LSPR offer complementary advantages. The combination of the photophysical properties of QDs with the enhanced sensitivity of LSPR can lead to more robust biosensors, imaging agents, and therapeutic systems. For example, QD-LSPR hybrids can significantly improve the specificity and sensitivity of disease detection, such as in cancer diagnostics, where early detection is crucial.

Overall, the integration of quantum dots with localized surface plasmon resonance offers substantial potential for advancing biomedical applications. This synergy holds promise for improving diagnostic accuracy, targeted therapy, and real-time monitoring of disease progression, making them invaluable in the future of personalized medicine and healthcare.

Acknowledgements All the authors thank Manipal College of Pharmaceutical Sciences, MAHE, for supporting this work.

References

1. Wagner, A. M., Knipe, J. M., Orive, G., & Peppas, N. A. (2019). Quantum dots in biomedical applications. *Acta Biomaterialia, 94*, 44–63. https://doi.org/10.1016/j.actbio.2019.05.022
2. Mcoyi, M. P., Mpofu, K. T., Sekhwama, M., & Mthunzi-Kufa, P. (2024). Developments in localized surface Plasmon resonance. *Plasmonics, 1–40*. https://doi.org/10.1007/s11468-024-02620-x
3. Long, Y.-T., & Jing, C. (2024). *Localized surface plasmon resonance based nanobiosensors*. http://www.springer.com/series/8898
4. Michalet, X., Pinaud, F. F., Bentolila, L. A., Tsay, J. M., Doose, S., Li, J. J., Sundaresan, G., Wu, A. M., Gambhir, S. S., & Weiss, S. (2005). Quantum dots for live cells, in vivo imaging, and diagnostics. *Science, 307*, 538–544. www.sciencemag.org/cgi/content/full/307/5709/538/DC1
5. Rizvi, S. B., Ghaderi, S., Keshtgar, M., & Seifalian, A. M. (2010). Semiconductor quantum dots as fluorescent probes for in vitro and in vivo biomolecular and cellular imaging. *Nano Reviews, 1*(1), 1–15. https://doi.org/10.3402/nano.v1i0.5161
6. Chang, Y. P., Pinaud, F., Antelman, J., & Weiss, S. (2008). Tracking biomolecules in live cells using quantum dots. *Journal of Biophotonics, 1*(4), 287–298. https://doi.org/10.1002/jbio.200810029
7. Matea, C. T., Mocan, T., Tabaran, F., Pop, T., Mosteanu, O., Puia, C., Iancu, C., & Mocan, L. (2017). Quantum dots in imaging, drug delivery and sensor applications. *International Journal of Nanomedicine, 12*, 5421–5431. https://doi.org/10.2147/IJN.S138624
8. Yin, C., Zhu, Y., Lv, Y., Deng, H., & Liu, B. (2023). The potential of engineered multifunctional quantum dots for macrophage theranostics. *Innovations, 4*(5), 1–2. https://doi.org/10.1016/j.xinn.2023.100492
9. Hong, Y., Ku, M., Lee, E., Suh, J.-S., Huh, Y.-M., Yoon, D. S., & Yang, J. (2013). Localized surface plasmon resonance based nanobiosensor for biomarker detection of invasive cancer cells. *Journal of Biomedical Optics, 19*(5), 1–7. https://doi.org/10.1117/1.jbo.19.5.051202

10. Lan, Y.-B., Wang, S.-Z., Yin, Y.-G., Hoffmann, W. C., & Zheng, X.-Z. (2008). Using a surface Plasmon resonance biosensor for rapid detection of salmonella typhimurium in chicken carcass. *Journal of Bionic Engineering, 5*, 239–246.
11. Nguyen, H. H., Park, J., Kang, S., & Kim, M. (2015). Surface plasmon resonance: A versatile technique for biosensor applications. *Sensors (Switzerland), 15*(5), 10481–10510. https://doi.org/10.3390/s150510481
12. Hao, X., St-Pierre, J. P., Zou, S., & Cao, X. (2023). Localized surface plasmon resonance biosensor chip surface modification and signal amplifications toward rapid and sensitive detection of COVID-19 infections. *Biosensors and Bioelectronics, 236*, 1–9. https://doi.org/10.1016/j.bios.2023.115421
13. Zhang, P., Chen, Y. P., Wang, W., Shen, Y., & Guo, J. S. (2016). Surface plasmon resonance for water pollutant detection and water process analysis. *TrAC - Trends in Analytical Chemistry, 85*, 153–165. https://doi.org/10.1016/j.trac.2016.09.003
14. Nnachi, R. C., Sui, N., Ke, B., Luo, Z., Bhalla, N., He, D., & Yang, Z. (2022). Biosensors for rapid detection of bacterial pathogens in water, food and environment. *Environment International, 166*, 1–20. https://doi.org/10.1016/j.envint.2022.107357
15. Das, S., Devireddy, R., & Gartia, M. R. (2023). Surface Plasmon Resonance (SPR) sensor for cancer biomarker detection. *Biosensors, 13*(3), 1–33. https://doi.org/10.3390/bios13030396
16. Nurrohman, D. T., Chiu, N. F., Hsiao, Y. S., Lai, Y. J., & Nanda, H. S. (2024). Advances in nanoplasmonic biosensors: Optimizing performance for exosome detection applications. *Biosensors, 14*(6), 1–31. https://doi.org/10.3390/bios14060307
17. Bera, D., Qian, L., Tseng, T. K., & Holloway, P. H. (2010). Quantum dots and their multimodal applications: A review. *Materials, 3*(4), 2260–2345. https://doi.org/10.3390/ma3042260
18. Jain, P. K., Eustis, S., & El-Sayed, M. A. (2006). Plasmon coupling in nanorod assemblies: Optical absorption, discrete dipole approximation simulation, and exciton-coupling model. *Journal of Physical Chemistry B, 110*(37), 18243–18253. https://doi.org/10.1021/jp063879z
19. Varadharajan, S., Vasanthan, K. S., Mathur, V., Hariperumal, N., & Mazumder, N. (2024). Green synthesis and multifaceted applications: Challenges and innovations in carbon dot nanocomposites. *Discover Nano, 19*(1), 1–47. https://doi.org/10.1186/s11671-024-04124-3

Cadmium Telluride Quantum Dots for Cancer Diagnosis

Anagha Ajjanagadde, Medha Bairy, Rajat Hegde, and Shama Prasada Kabekkodu

1 Introduction

Quantum dots (QDs) are nanoscale semiconductors and fluorophores with unique optical properties and are confined within 100–1000 atoms in three spatial dimensions, ranging from 1.5–10 nm in size [1]. Quantum dots are useful in numerous fields because of their unique features, such as high quantum efficiency, fluorescence lifespan, constrained emission spectra, biological compatibility, and photothermal and photochemical stability [2]. The spectral characteristics of cadmium telluride quantum dots (CdTe QDs), which are composed of cadmium telluride atoms, include broad absorption, limited emission, and photostability. These characteristics make them highly promising for therapeutic applications and medical imaging applications. CdTe QDs are of great interest for their prospective uses in gamma ray detectors, sensors, solar power cells, optical electronics, biolabelling, and biological imaging [3]. CdTe QDs have drawn much scientific interest recently because of their high quantum yield, strong fluorescence efficiency, and adjustable bandwidth. By adjusting their particle size, they can exhibit adjustable photoluminescence characteristics in the visible band (~500–750 nm). Furthermore, all colors of CdTe QDs can be excited simultaneously by a single excitation wavelength of light [4].

A. Ajjanagadde · M. Bairy · R. Hegde · S. P. Kabekkodu (✉)
Department of Cell and Molecular Biology, Manipal School of Life Sciences, Manipal Academy of Higher Education, Manipal, Karnataka, India
e-mail: shama.prasada@manipal.edu

S. P. Kabekkodu et al. (eds.), *Functional Quantum Dots*, Lecture Notes in Nanoscale Science and Technology 40,
https://doi.org/10.1007/978-3-031-98219-4_6

2 Key Features of Cadmium Telluride Quantum Dots

Quantum dots are semiconductor components that are usually between 2 and 20 nm in size and are made of binary and ternary alloys from periodic table categories II-VI, III-V, and IV-VI [5]. The primary advantages are as follows.

- Comparatively prolonged fluorescence lifetime with excellent optical stability.
- The capacity to excite several QDs at once with a single light.
- Narrow emission spectra that can be tuned with broad absorption spectra.
- Size-dependent emission and high quantum yield.
- Strong photostability [6].

In general, smaller QDs exhibit higher energy, higher frequency, and a shift toward blue light emission. CdTe QDs exhibit size-dependent optical properties due to the quantum confinement effect, making them versatile in bioimaging and opto-electronics applications because of their size-dependent nature [7]. Some of the key features are as follows:

2.1 Quantum Confinement Effect

Since CdTe quantum dots are smaller than the Bohr exciton radius, the quantum confinement effect alters both their electrical and optical characteristics [8]. The quantum dots in the CdTe QDs exhibit a blueshift in the absorption and emission spectra due to a size-dependent band gap, which enhances the electron–hole interaction and photoluminescence efficiency [9].

2.2 High Fluorescence

CdTe QDs display intense fluorescence due to their direct bandgap structure, allowing efficient radiative recombination of electron–hole pairs, resulting in strong emission upon photon absorption. Because of the quantum confinement effect, larger quantum dots emit low-energy red light, and smaller quantum dots emit high-energy blue light, making the fluorescence extremely flexible. The high fluorescence of CdTe QDs is due to their high extinction coefficient, surface passivation techniques, and core–shell structures, which enhance fluorescence by reducing non-radiative recombination [10]. Cadmium telluride quantum dots (CdTe QDs) are utilized extensively in bioimaging, labeling, biosensors, LEDs, and photodetectors due to their brilliant, stable fluorescence and their adaptability in a wide range of applications.

2.3 High Photostability

CdTe QDs offer photostability, surpassing traditional organic dyes, making them crucial for applications such as live-cell imaging and optical devices requiring sustained fluorescence intensity. The improved photostability of the CdTe QDs can be attributed to several factors. CdTe QDs with zinc sulfide (ZnS) encapsulation enhance photostability and resistance to photobleaching, forming core–shell structures that reduce surface defects and prevent oxidative degradation [11]. The photostability of CdTe QDs is influenced by size, with larger QDs exhibiting greater photostability, which is crucial for specific applications, especially under continuous illumination. The choice of capping ligands significantly influences the photostability of CdTe QDs, with certain ligands enhancing their stability and photoresponsivity, making them ideal for biological applications requiring prolonged fluorescence [12].

2.4 Broad Absorption Spectrum

CdTe QDs are versatile because of their broad absorption spectrum, making them ideal for various applications. The absorption range of 300–800 nm allows for simultaneous excitation of multiple QDs with varying emission wavelengths via a single light source, enabling multiplexed imaging and sensing applications. The high molar extinction coefficients and size-dependent energy levels of CdTe QDs enable efficient light absorption across a wide spectral range, making them effective in biosensors and photovoltaic applications [13].

2.5 High Quantum Yield

CdTe QDs are known for their high photoluminescence quantum yield, often exceeding 70%, indicating their ability to convert absorbed photons into emitted fluorescence. An elevated quantum yield is achieved through strategic surface passivation and core–shell engineering techniques, which mitigate surface defects and reduce nonradiative recombination pathways in the QD core. Encapsulating CdTe cores with zinc or cadmium sulfide shells enhances surface quality by acting as a protective barrier, reducing defects and preventing oxidative degradation [14]. Owing to their ability to fine-tune the quantum yield through surface modifications and core–shell engineering, CdTe QDs are highly desirable for applications in bioimaging, sensing, and optoelectronic devices.

2.6 Size Adjustable Emission

The quantum confinement phenomenon of CdTe QDs allows for size-adjustable emission properties, which allows for precise control throughout the optical characteristics of the particles during synthesis. The size of the CdTe QDs decreases, causing discrete energy levels and increased bandgap energy, causing emission wavelength shifts [15]. This tunability is due to the quantum confinement effect, which affects the electronic and optical properties. The precise control of synthesis parameters such as the reaction time, temperature, and precursor concentration enables the fabrication of CdTe QDs with specific sizes and emission properties.

3 Methods of CdTe QD Synthesis

CdTe QDs are synthesized via various methods to achieve specific properties for diverse applications, and the primary synthesis techniques include the following:

3.1 Aqueous Phase Synthesis

The method involves the production of CdTe QDs in water, which is simple, cost-effective, and environmentally friendly. Cadmium precursors such as $CdCl_2$ typically react with tellurium sources, with stabilizing agents such as thioglycolic acid or mercaptopropionic acid. Stabilizers are used to cover the QD surfaces, preventing aggregation and controlling growth. The precise control of reaction conditions, such as temperature and pH, is crucial for achieving the desired QD size and optical properties. CdTe QDs, synthesized using water as a solvent, have been successfully produced under mild conditions at temperatures of 70, 80, and 90 °C [16]. This technique yields QDs with quantum yields of up to 80%, making them ideal for biophotonic applications.

3.2 Organic Phase Synthesis

The method involves synthesizing CdTe QDs in organic solvents at high temperatures, often exceeding 200 °C. The reaction medium is typically high-boiling-point solvents such as TOPO or ODE. High temperatures cause the decomposition of cadmium and tellurium precursors, resulting in the nucleation and growth of quantum dots. Organic ligands such as oleic acid or trioctylphosphine (TOP) stabilize QDs, affecting their size and optical properties [17]. This method produces high-quality QDs with narrow size distributions, but hydrophobic nanoparticles require surface modification for aqueous applications.

3.3 Green Synthesis Method

Synthesis routes have been developed with an environmental focus to minimize toxicity and environmental impact. The method involves combining plant-mediated techniques with microwave and ultrasonic-assisted extraction. This technique uses natural plant extracts as reducing and capping agents, enabling the rapid and environmentally friendly synthesis of CdTe QDs. The process is in line with green chemistry principles and offers benefits such as reduced reaction times and energy consumption [18].

3.4 Single-Source Precursor Approach

This method utilizes single-molecule precursors containing both cadmium and tellurium, simplifying the synthesis process. $Cd(TePh)_2$, a compound, can be decomposed via the use of amine solvents and capping agents to produce CdTe QDs, increasing the degree of stoichiometry control and achieving the desired properties [19].

3.5 Core–Shell Structure Formation

Core–shell structures are frequently synthesized to improve the optical properties and stability of CdTe QDs. The process involves the formation of a shell of another semiconductor material, such as zinc sulfide (ZnS) or cadmium sulfide (CdS), around the CdTe cores. The shell effectively passivates surface defects, reduces nonradiative recombination, and enhances the quantum yield and photostability. For example, CdTe/ZnS core/shell QDs have been synthesized using molecular precursors, leading to enhanced optical properties [20]. A schematic representation of the chemical and biological synthesis of QDs is shown in Fig. 1.

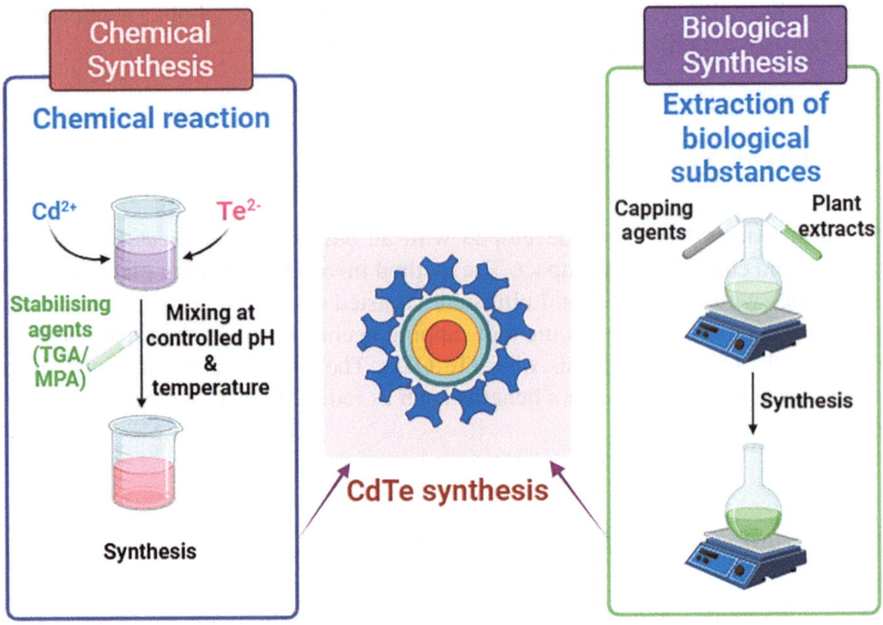

Fig. 1 Illustration of CdTe QD synthesis

4 Cadmium Telluride Quantum Dot Applications in Oncology

4.1 Detection of Cancer and Imaging

CdTe QDs are highly effective in cancer diagnostics because of their fluorescence properties and high sensitivity. CdTe QDs coupled with the nucleolin-targeting aptamer AS1411 and plated onto silica nanoparticles have been investigated to create a highly selective dual aptamer bioprobe. With only 85 cells per milliliter, the probe showed remarkable efficacy in identifying MCF-7 breast cancer cells. The excellent selectivity of the QD-based aptasensor makes it a promising technique for detecting breast cancer in its early stages, which could enhance screening and diagnosis [21].

Breast cancer is a prevalent global disease, with sentinel lymph node (SLN) biopsy being a crucial procedure for cancer staging and metastasis assessment. Conventional SLN mapping techniques, which rely on organic dyes and radiocolloids, can lead to tissue damage and adverse effects. Quantum dots provide a noninvasive, high-resolution imaging method with enhanced sensitivity. Compared with traditional methods, functionalized QDs enable real-time visualization of sentinel lymph nodes, enhancing tumor staging accuracy and reducing complications [22].

QD integration in image-guided tumor resection enhances surgical precision, improves cancer cell visualization, and reduces tumor recurrence by ensuring the complete removal of malignant tissue. Despite their potential, the clinical application of QDs faces challenges related to their biocompatibility and toxicity. Research has focused on optimizing the surface chemistry of CdTe QDs to increase their stability and reduce their cytotoxic effects. This includes developing biocompatible coatings and functionalized QDs that maintain fluorescence efficiency while minimizing adverse interactions with biological systems, potentially leading to the clinical adoption of QD-based imaging and diagnostic technologies.

4.2 *In Vitro and In Vivo Imaging*

MSA-capped CdTe/anti-GFAP QDs have shown potential for tumor labelling in both in vitro and in vivo models, enabling precise visualization of tumor cells for guidance of intraoperative tumor resection. The application of these methods in imaging-guided surgery and monitoring three-dimensional glial cultures suggests their broader utility in neuro-oncological research [23].

Quantum dots with near-infrared emission are highly desirable for their superior photostability, fluorescence efficiency, and tunable optical properties, making them ideal for bioimaging and biosensing applications. This study evaluated the intracellular photostability of CdHgTe/CdS/CdZnS QDs coated with NAC, MPA, and TGA ligands in HeLa cells [24]. NAC-capped quantum dots exhibited fluorescence endurance under UV light, demonstrating the function of NAC in halting QD degradation and implying that their fluorescence durability makes them potentially trustworthy probes for long-term cellular imaging [25].

Another study used NIR CdTe/CdS QDs stabilized with MPA as nanosensors to monitor the Cu^{2+} concentration in living cells. HeLa cells presented strong fluorescence signals, confirming efficient uptake. However, pretreatment with Cu^{2+} caused fluorescence quenching (~90%) due to QD aggregation. This study highlights the high selectivity, photostability, and rapid response of CdTe-based QD nanosensors, demonstrating their potential for real-time imaging and metal ion detection in biomedical applications [26].

CdTe-based QDs are crucial in cancer diagnostics and therapeutic strategies, and further research and optimization can increase their clinical translation for improved precision in oncology, highlighting their versatility in tumor imaging and biosensing.

4.3 *Multiplexed Detection of Cancer Biomarkers*

The development of sensitive and specific cancer biomarker detection tools, such as CdTe/CdS QDs, is crucial for early diagnosis and effective clinical management, offering superior sensitivity and specificity in cancer diagnostics. Aqueous-phase-synthesized CdTe/CdS QDs have been used to create a microfluidic protein chip that can identify cancer biomarkers. As luminous signal amplifiers, these QDs increase the sensitivity of detection. By conjugating secondary antibodies to quantum dots, many cancer biomarkers can be detected simultaneously, offering a high-throughput, sensitive, and specific diagnostic device [27].

Quantum dot-based immunohistofluorescence technology has been utilized to assess the subcellular expression levels of multiplexed biomarkers, enhancing the predictive accuracy for lymph node metastasis, a crucial factor in determining prognosis and treatment strategies [28]. Compared with single biomarker assessments, the combination of EGFR and E-cadherin showed superior predictive power for LNM, DFS, and OS in HNSCC patients [28]. Multiplexing is advantageous for limited tissue samples, as it allows simultaneous evaluation of multiple biomarkers within the same sample. The integration of quantum dot-based biomarker multiplexing in clinical diagnostics could improve cancer prognosis and guide personalized treatment strategies. This study demonstrates the potential of QD technologies in establishing robust predictive models for cancer metastasis and patient outcomes. Further validation studies incorporating additional biomarkers and clinical factors are needed to refine this approach for routine clinical application.

4.4 *Drug Delivery*

According to investigations, nanoparticles can improve bioavailability, decrease toxicity, regulate release, and increase drug circulation. Doxorubicin was loaded into PEG-CdTe QDs for targeted drug delivery, promoting intracellular drug accumulation, pH-controlled release, and enhanced therapeutic efficacy by increasing apoptosis in myeloma cells. This system could be a potential clinical treatment for EMD patients [29]. To enhance the fluorescence characteristics and doxorubicin loading efficiency, CdTe QDs were added to the hollowed mesoporous silica nanoparticles (hMSNs). This hybrid system preserved fluorescence while facilitating tumor-targeted drug delivery and accumulation. Functionalization with VEGF antibodies enhances the effectiveness of these conjugates, making them promising candidates for cancer detection and therapy [30]. Schematic representation of the applications of CdTe QDs in oncology is shown in Fig. 2.

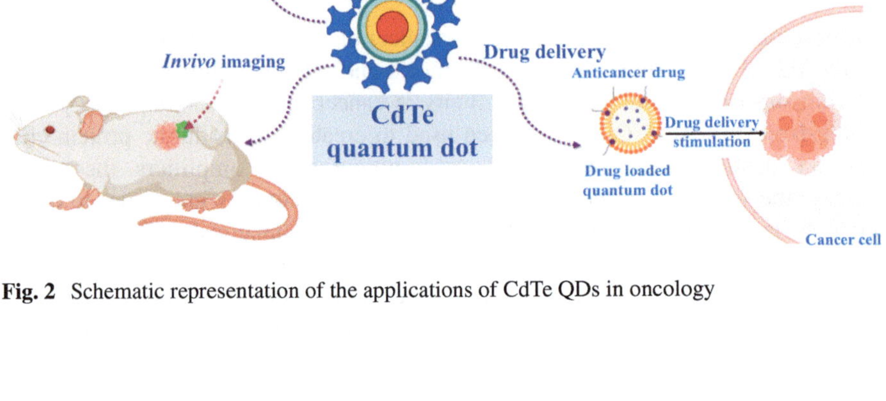

Fig. 2 Schematic representation of the applications of CdTe QDs in oncology

5 Conclusion

Recently, quantum dots have emerged as nanoparticles with special qualities that may influence the diagnosis and treatment of cancer. Extremely large surface-to-volume ratios, stable photoluminescence, small adaptable sizes, and possible biocompatibility are some of these characteristics. QDs can be bioconjugated or have their surface modified to suit the application, which makes them viable options for multifunctional systems. Owing to their strong fluorescence, QDs are useful for imaging, biomarker tagging, and molecular marker detection as signal reporters in cancer diagnosis. These methods may increase the precision of early cancer detection through the use of sensitive analysis of tumor-specific markers and high-resolution imaging of tumor tissues. QDs have many potential uses in cancer treatment. When coupled with different biomolecules, they can be utilized for controlled release, intraoperative navigation, and as drug carriers for accurate medication administration. Studies indicate that QDs may be useful in cancer immunotherapy, which could result in new approaches to cancer treatment. In conclusion, even though QDs present promising opportunities for cancer detection, treatment, and prevention, efforts focused on biosafety, biocompatibility, and translational research are needed for their clinical use. QDs have the potential to significantly contribute to the advancement of cancer research and patient care with further innovation and multidisciplinary collaboration.

Acknowledgement The authors would like to acknowledge the Wellcome Trust DBT India Alliance Intermediate Fellowship (IA/I/22/1/506240), SPARC-Ministry of Human Resource Development, Government of India (P2297), SERB-MATRICS, Government of India. (MTR/2021/000182). The authors also thank the Manipal School of Life Sciences, Manipal Academy of Higher Education, Manipal, India.

References

1. Singh, S., Dhawan, A., Karhana, S., Bhat, M., & Dinda, A. K. (2020). Quantum dots: An emerging tool for point-of-care testing. *Micromachines, 11*(12), 1–23. https://doi.org/10.3390/mi11121058
2. Pandey, S., Mukherjee, D., Kshirsagar, P., Patra, C., & Bodas, D. (2021). Multiplexed bioimaging using cadmium telluride quantum dots synthesized by mathematically derived process parameters in a continuous flow active microreactor. *Materials Today, Bio, 11*. https://doi.org/10.1016/j.mtbio.2021.100123
3. Mondal, J., Lamba, R., Yukta, Y., Yadav, R., Kumar, R., Pani, B., & Singh, B. (2024). Advancements in semiconductor quantum dots: Expanding frontiers in optoelectronics, analytical sensing, biomedicine, and catalysis. *Journal of Materials Chemistry C, 12*(28), 10330–10389. https://doi.org/10.1039/d4tc01396d
4. Borah, D. J., Saikia, D., Das, A., Saikia, P. K., & Mostako, A. T. T. (2023). An investigation on the structural and optical properties of mercaptosuccinic acid capped cadmium telluride quantum dots. *Discover Materials, 3*(1). https://doi.org/10.1007/s43939-023-00041-x

5. Panja, A., & Patra, P. (2023). A review on quantum dots (QDs) and their biomedical applications. *4open, 6*, 1. https://doi.org/10.1051/fopen/2022020
6. Barroso, M. M. (2011). Quantum dots in cell biology. *Journal of Histochemistry and Cytochemistry, 59*(3), 237–251. https://doi.org/10.1369/0022155411398487
7. Bera, D., Qian, L., Tseng, T. K., & Holloway, P. H. (2010). Quantum dots and their multimodal applications: A review. *Materials, 3*(4), 2260–2345). MDPI AG. https://doi.org/10.3390/ma3042260
8. Kalsoom, U. E., Yi, R., Qu, J., & Liu, L. (2021). Nonlinear optical properties of CdSe and CdTe Core-Shell quantum dots and their applications. *Frontiers in Physics, 9*. https://doi.org/10.3389/fphy.2021.612070
9. Sanmartín-Matalobos, J., Bermejo-Barrera, P., Aboal-Somoza, M., Fondo, M., García-Deibe, A. M., Corredoira-Vázquez, J., & Alves-Iglesias, Y. (2022). Semiconductor quantum dots as target analytes: Properties, surface chemistry and detection. *Nanomaterials, 12*, 14). MDPI. https://doi.org/10.3390/nano12142501
10. Rizvi, S. B., Ghaderi, S., Keshtgar, M., & Seifalian, A. M. (2010). Semiconductor quantum dots as fluorescent probes for in vitro and in vivo biomolecular and cellular imaging. *Nano Reviews, 1*(1), 5161. https://doi.org/10.3402/nano.v1i0.5161
11. Wang, X., Wang, P., Li, M., & Li, J. (2024). Advances in the preparation and biological applications of core@shell nanocrystals based on quantum dots and noble metal. *RSC Advances, 14*(36), 26308–26324. https://doi.org/10.1039/d4ra05386a
12. Huang, S., & Huang, G. (2024). The utilization of quantum dot labelling as a burgeoning technique in the field of biological imaging. *RSC Advances, 14*(29), 20884–20897. https://doi.org/10.1039/d4ra04402a
13. Shen, X., Jia, J., Lin, Y., & Zhou, X. (2015). Enhanced performance of CdTe quantum dot sensitized solar cell via anion exchanges. *Journal of Power Sources, 277*, 215–221. https://doi.org/10.1016/j.jpowsour.2014.12.022
14. Martins, C. S. M., Silva, A. L., de Gouveia, L. P., Çaha, I., Bondarchuk, O., LaGrow, A. P., Deepak, F. L., & Prior, J. A. V. (2024). Controlling fluorescence wavelength in the synthesis of TGA-capped CdTe quantum dots. *Chem, 12*, 4. https://doi.org/10.3390/chemosensors12040070
15. Pandey, S., & Bodas, D. (2023). Exploiting the UV excited size-dependent emission of PDMS-coated CdTe quantum dots for in vitro simultaneous multicolor imaging of HepG2 cellular organelles. *Materials Advances, 4*(7), 1694–1701. https://doi.org/10.1039/d2ma00964a
16. Alper, F. M. P. (2023). Aqueous synthesis of CdTe quantum dots and optical characterization under high voltage. *Acta Physica Polonica A, 143*(3), 221–227. https://doi.org/10.12693/APhysPolA.143.221
17. Farkhani, S. M., & Valizadeh, A. (2014). Review: Three synthesis methods of CdX (X = Se, S or Te) quantum dots. *IET Nanobiotechnology, 8*(2), 59–76. https://doi.org/10.1049/iet-nbt.2012.0028
18. Moradi Alvand, Z., Rajabi, H. R., Mirzaei, A., Masoumiasl, A., & Sadatfaraji, H. (2019). Rapid and green synthesis of cadmium telluride quantum dots with low toxicity based on a plant-mediated approach after microwave and ultrasonic assisted extraction: Synthesis, characterization, biological potentials and comparison study. *Materials Science and Engineering C, 98*, 535–544. https://doi.org/10.1016/j.msec.2019.01.010
19. Taniguchi, S., & Green, M. (2015). The synthesis of CdTe/ZnS core/shell quantum dots using molecular single-source precursors. *Journal of Materials Chemistry C, 3*(32), 8425–8433. https://doi.org/10.1039/c5tc01808k
20. Haque, M., Chutia, J., Mondal, A., Quraishi, S., Kumari, K., Marboh, E. W. M., Aguan, K., & Singha Roy, A. (2024). Formation of CdTe core and CdTe@ZnTe core-shell quantum dots via hydrothermal approach using dual capping agents: Deciphering the food dye sensing and protein binding applications. *Physical Chemistry Chemical Physics*. https://doi.org/10.1039/d4cp02225d
21. Omidian, H., Wilson, R. L., & Cubeddu, L. X. (2024). Quantum dot research in breast cancer: Challenges and prospects. *Materials, 17*, 9. https://doi.org/10.3390/ma17092152

22. Alam, F., & Yadav, N. (2013). Potential applications of quantum dots in mapping sentinel lymph node and detection of micrometastases in breast carcinoma. *Journal of Breast Cancer, 16*(1), 1–11. https://doi.org/10.4048/jbc.2013.16.1.1
23. Seabra, M. A. B. L., Dubois, L. G., dos Santos-Júnior, E. F., Cavalcanti Santos, R. V., de Castro Neto, A. G., Isaac, A. R., Fontes, A., Hochhaus, G., da Costa, B. L. D. S. A., Moura Neto, V., & Santos, B. S. (2024). CdTe quantum dots as fluorescent nanotools for in vivo glioblastoma imaging. *Optical Materials: X, 21*. https://doi.org/10.1016/j.omx.2023.100282
24. Liu, X., Zhou, P., Liu, H., Zhan, H., Zhang, Q., Zhao, Y., & Chen, Y. (2018). Design of bright near-infrared-emitting quantum dots capped with different stabilizing ligands for tumor targeting. *RSC Advances, 8*(8), 4221–4229. https://doi.org/10.1039/c7ra10824a
25. Dan, Z., Zhike, H., Chan, W. H., & Martin, M. F. C. (2009). Synthesis and characterization of high-quality water-soluble near-infrared-emitting CdTe/CdS quantum dots capped by N-acetyl-L-cysteine via hydrothermal method. *Journal of Physical Chemistry C, 113*(4), 1293–1300. https://doi.org/10.1021/jp808465s
26. Hamidu, A., Pitt, W. G., & Husseini, G. A. (2023). Recent breakthroughs in using quantum dots for cancer imaging and drug delivery purposes. *Nanomaterials, 13*, 18. https://doi.org/10.3390/nano13182566
27. Hu, M., Yan, J., He, Y., Lu, H., Weng, L., Song, S., Fan, C., & Wang, L. (2010). Ultrasensitive, multiplexed detection of cancer biomarkers directly in serum by using a quantum dot-based microfluidic protein chip. *ACS Nano, 4*(1), 488–494. https://doi.org/10.1021/nn901404h
28. Hu, Z., Qian, G., Müller, S., Xu, J., Saba, N. F., Kim, S., Chen, Z., Jiang, N., Wang, D., Zhang, H., Lane, K., Hoyt, C., Shin, D. M., & Chen, Z. G. (2016). Biomarker quantification by multiplexed quantum dot technology for predicting lymph node metastasis and prognosis in head and neck cancer. *Oncotarget, 7*(28), 44676–44685. https://doi.org/10.18632/oncotarget.9225
29. Chen, D., Chen, B., & Yao, F. (2018). Doxorubicin-loaded PEG-CdTe quantum dots as a smart drug delivery system for extramedullary multiple myeloma treatment. *Nanoscale Research Letters, 13*. https://doi.org/10.1186/s11671-018-2782-0
30. Yang, D., Wang, N., Ji, H., Sun, S., Dong, J., Zhong, Y., Qian, C., & Xu, H. (2018). Preparation of core/shell CdTe@hMSN for enhanced tumor vasculature-specific drug delivery. *RSC Advances, 8*(68), 38987–38994. https://doi.org/10.1039/c8ra07193d

Advanced Optical Microscopy for Oral Cancer Detection Using Quantum Dots

Doyal Dey, Arushi Kapoor, Rajib Biswas, Alevoor Srinivas Bharath Prasad, and Nirmal Mazumder

1 Introduction to QDs

Nanooncology focuses on nanotechnology in cancer diagnosis and treatment [1]. Owing to their minute size and specificity, nanomaterials are used in oncology [2]. Semiconductor QDs have intense and stable fluorescence; owing to this property, semiconductor QDs are used to detect cancer tissue biopsies and are also used in medical imaging [3]. QDs are semiconductor nanocrystals that possess quantum confinement properties. When these QDs are excited, they emit fluorescence between the visible spectrum and the infrared region. Each crystal unit of a quantum dot contains 100–10,000 atoms, and the size ranges from 2 to 10 nm [4]. QDs can be divided into categories on the basis of their placement in the periodic table. The first category is from Groups III—V of the periodic table, whereas the second category is from Groups II–VI. The second category is more hazardous and releases heavy elements. Cadmium selenide (CdSe) is a widely used material in QDs. The outer shell of these QDs is composed of zinc sulfide.

D. Dey · A. S. B. Prasad
Department of Public Health Genomics, Manipal School of Life Sciences, Manipal Academy of Higher Education, Manipal, Karnataka, India

A. Kapoor
School of Civil & Environmental Engineering and Construction Management, University of Texas at San Antonio, San Antonio, TX, USA

R. Biswas
Department of Physics, Tezpur University, Tezpur, Assam, India

N. Mazumder (✉)
Department of Biophysics, Manipal School of Life Sciences, Manipal Academy of Higher Education, Manipal, Karnataka, India
e-mail: nirmal.mazumder@manipal.edu

S. P. Kabekkodu et al. (eds.), *Functional Quantum Dots*, Lecture Notes in Nanoscale Science and Technology 40,
https://doi.org/10.1007/978-3-031-98219-4_7

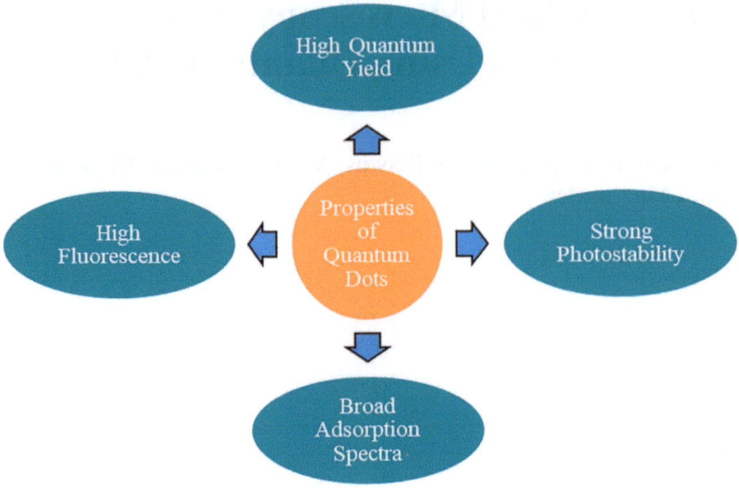

Fig. 1 Properties of QDs

QDs have properties such as high fluorescence, broad absorption spectra, strong photostability, and size-tunable emission, as shown in Fig. 1. The optoelectronic properties of QDs are due to the quantum confinement effect [5].

Many conventional imaging techniques face challenges due to the lack of fluorescent markers. QDs provide a promising solution to this problem because of the photophysical properties of fluorophore nanoparticles for in vivo fluorescence imaging. QDs can be used as sensitive probes for screening cancer markers in fluid samples. These fluorescent nanoparticles can be used to classify tissue biopsies and act as high-resolution contrast agents for medical testing, thus increasing their sensitivity for detecting tumors. Owing to their multiplexing capabilities, QDs can detect many cancer markers simultaneously.

1.1 Overview of Oral Cancer Treatment

Cancer is a disease in which there is an uncontrolled proliferation of abnormal cells; these cells invade the tissues in the body and metastasize. Every year, approximately 4,00,000 cases of oral cancer are estimated worldwide, with two-thirds occurring in countries such as India, Pakistan, Bangladesh, and Sri Lanka. Oral cancer is the sixth most common cancer malignancy worldwide [4]. Oral squamous cell carcinomas (OSCCs) account for approximately 90% of all reported oral cancer cases. These tumors can spread to other parts of the mouth, such as the buccal, tongue alveolar, hard palate, and tongue. Some of the risk factors leading to oral cancer include smoking, human papillomavirus infection, and the accumulation of genetic mutations. Oral cancer has a 5-year survival rate of approximately 40%. Easy detection significantly improves outcomes, with survival rates of 80% for cancer patients diagnosed at the first and second stages; however, due to the poor prognosis of oral cancer patients, approximately half of the oral cancer patients are diagnosed at the advanced stages of the third and fourth stages [6]. A delay in the diagnosis of the disease could lead to an increased risk of having an advanced stage of cancer [4].

1.2 Advantages of QD Imaging

The gold standard for the detection of oral cancer has been conventional oral examination involving histopathological analysis, which lacks accuracy in the detection of molecular changes. The properties of QDs include resistance to photobleaching, size-tunable light emission, and good fluorescence intensity. QDs can be excited via a single wavelength, and the ability of QDs to link with peptides and antibodies can be optimized. Compared with conventional optical imaging tests, QDs have greater potential for degradation. QDs are photostable, and when exposed to light for more than 1 h, the QDs still exhibit constant excitation. The single excitation source of the quantum points makes QDs more efficient. QDs can be used as existing excitation sources since they have a broad excitation spectrum [7]. QDs have several advantages because the nanometer-sized semiconductor crystals that make up the structure of the QDs. These crystals produce luminescence due to the phenomenon known as the quantum confinement effect. Owing to this effect, the interaction of the QDs with light causes the QDs to emit light. Some of the advantages of using QDs in comparison with other fluorescent proteins and organic dyes are that QDs have unique optical and electrical properties for use in cellular imaging. Some other advantages of QDs include broad light spectrum emission from the visible to the infrared range due to their size, photostable fluorescence, and water solubility [8]. QDs play an important role in tumor imaging and are used in near-infrared (NIR, 700–900 nm) imaging. NIR provides a clear image for in vivo optical imaging of the tumor [9]. QDs also have the potential to be used for preventative diagnosis in oral oncology. QDs can be put into biosensors and can be used for the identification of cancer biomarkers as well as genetic mutations. The use of these biosensors can aid in the early detection of oral cancer. QDs can identify different cancer biomarkers at once because of their adjustable emission wavelength. This offers sensitive and focused tests for early cancer detection [10, 11]. QDs can also be used for genetic screening assays. These assays can help in the early detection of the risk of a person having cancer. QDs can be used in SNP genotyping assays. These assays can help people who are at high risk of cancer, as this information can help in the early treatment of the disease [12].

2 Classification of Different Types of QDs

The core type of QDs containing selenium is the most studied type of QD.

2.1 Cadmium Selenide QDs (CdSe QDs)

CdSe QDs are spherical-shaped nanoparticles ranging from 1 to 12 nm in size. Biologically synthesized CdSe shows good water solubility and has an excitation wavelength of 295 nm. CdSe QDs have unique properties, such as biocompatibility and visible optical and electrical properties. Various applications of CdSe QDs include biomolecular tracking, binding assays, and in vitro imaging, such as cellular imaging and tumor imaging [13]. The many applications of QDs include protein tracking, drug delivery, disease diagnosis, and intracellular imaging. Owing to the emission characteristics of QDs, these nanoparticles can be used along with drug carriers and can also be used as drug delivery systems [14]. CdSe QDs are light-emitting nanoparticles with photophysical properties that also allow them to be used for targeted drug delivery applications and biological analysis. QDs can penetrate many biomolecules, such as proteins, antibodies, and small-molecule ligands [14].

2.2 Cadmium Telluride QDs (CdTe QDs)

Core–shell QDs are second-generation QDs that were developed because of their ability to adjust the photophysical properties of the QDs. The shells of these QDs were designed to enhance their photostability and photoluminescence efficacy. The latest generation of highly luminescent QDs is alloyed QDs such as cadmium selenium sulfide (CdSeS) [13].

2.3 Carbon QDs (CQDs)

CQDs possess qualities such as high water solubility and tunable fluorescence properties. Carbon dots provide good photostability and can be synthesized cost-effectively. CQDs are good for bioimaging since they exhibit bright fluorescence. It has diverse applications in bioimaging, photocatalysis, drug delivery, and biosensors [15].

Table 1 Characteristics of different types of QDs used in biomedical imaging [18]

Name	Size	Application	Pros	Cons
Cd QDs	2–10 nm	Cd QDs are used in various bioimaging applications, such as tissue staining, binding assays, in vivo imaging, tumor imaging, and cellular imaging.	These QDs are well-studied and stable. They exhibit high quantum yield.	Cd QDs have limited biological applications due to the presence of cadmium (Cd) ions, resulting in toxicity concerns.
CQDs	Less than 10 nm	CQDs have multiple biomedical applications, such as drug delivery and gene therapy, and are used in vivo studies.	These QDs have minimum toxicity and are environmental friendly. They have in vivo applications and can be synthesized from waste.	Compared with the other QDs, these are relatively new and may exert minor toxicity when used in high concentrations.
In QDs	5–50 nm	It has a high potential for broader biomedical applications.	These QDs have the potential for broader biomedical applications compared to the Cd QDs.	These QDs can be moderately toxic.

2.4 Indium-Based QDs (in QDs)

Compared with cadmium-based QDs (Cd QDs), QDs exhibit weak acute local toxicity in vivo [16]. QDs are effective for bioimaging sentinel lymph nodes (SLNs) in breast cancer. Studies have also shown that the indium phosphide core zinc sulfide shell conjugated with anti-claudin 4 and anti-prostate stem cell antigens, which are pancreatic-specific monoclonal antibodies, is effective for the early detection of cancer [17] (Table 1).

3 QDs in Biomedical Applications

QDs have various applications in biomedicine and chemical biology. In the past few years, QDs have been used for cell tracking, genetic technology, fluorescent labelling of cellular proteins, and in vivo tumor biology and imaging [19].

3.1 Use of QDs for the Detection of Primary Tumors and Metastasis

QDs have been used for the detection of primary tumors and metastasis. During the transformation of regular cells into cancer cells, the protein expression profile changes. Proteins that are known as "cancer cell-specific antigens" are expressed on the surface of cells. MUCI (CD227) is expressed in more than 90% of human breast cancer cases [20]. These antigens can be used for targeted cancer immunotherapy. It can also help in the diagnosis of the disease, as it can act as a target marker for diagnosis. QDs that are conjugated with antibodies specific to these antigens can help in the identification of primary tumors.

3.2 Use of QDs for the Identification of Molecular Targets for Targeted Therapy

Many transmembrane receptor tyrosine kinases are also present and highly expressed on the surface of certain types of cancer cells. Tyrosine kinases can be used as molecular targets in targeted cancer therapy since they are involved in the proliferation and metastasis of cancer cells. Many inhibitors against these particular tyrosine kinase receptors are being studied clinically to help treat this disease. The conjugation of QDs with a particular antibody that is specific to the target or with a particular ligand is also effective for the detection of antigens. The conjugated QDs can be administered in vivo for active tumor targeting. Research studies involving the use of semiconductor QDs with HER-2 antibodies have aided in the detection of the breast cancer marker HER-2. These QDs that are conjugated with antibodies selectively bind only to HER-2-positive breast cancer cells in vitro. The ability of QDs to bind to specific receptors on tumor cells in vivo has yet to be tested [21].

Therapeutic antibodies target tyrosine receptors via targeted downregulation and inhibition of receptor function [22]. Measurement of the levels of receptors after antibody therapy is an excellent biodynamic marker that helps in monitoring the therapeutic efficacy of antibodies. In the future, QDs can be designed to measure receptor levels quantitatively after therapy. QDs that are covalently bonded with therapeutic agents can be used to evaluate the pharmacodynamics and delivery effectiveness of these agents in real time. Studies have shown that QDs conjugated with HER-2 siRNA are effective in tracking the delivery of the therapeutic siRNA, and they also help to monitor the effectiveness of the siRNA-mediated downregulation of the receptor in breast cancer cells [23].

3.3 Use of QDs for SLN Mapping

SLNs are lymph nodes that receive metastasizing cancer cells from a primary tumor. SLN biopsy is a procedure that helps in the removal of SLNs. This technique is a routine clinical procedure but is not widely used because it requires precise nodal mapping. Type II QDs can be used for SLN mapping since these types of QDs emit fluorescence in the NIR region. Research has shown that the presence of polydentate phosphine in QDs makes them soluble and stable in serum. Through transdermal injection, these QDs are rapidly localized to the SLN. QDs are effective for SLN mapping in different types of cancer [21].

4 Detection of QDs in Oral Cancer

The use of imaging probes for the detection of cancer antigens has become popular since this technique allows the identification of early-stage malignancies and metastases. QDs are used because they are known to provide precise molecular- and cellular-level diagnostics, and they also amplify ultrasensitive bioassays [25].

4.1 *In Vitro Detection Methods*

QDs are used for in vitro tumor imaging of cancer cells. Many studies have been performed to develop in vitro fluorescence images of malignant ovarian, breast, lung, hepatocellular, and pancreatic cells. Sulfur-doped carbon dots (S-CDs) have also been used effectively as photosensitizers for the detection of OSCC. These QDs were used as nanophotosensitizers for OSCC therapy; these QDs showed low cytotoxicity and were self-luminous. These QDs can also be used as effective nanoweapons for anticancer therapy, as shown in Fig. 2 [26]. Research has demonstrated that fluorescent nitrogen-doped carbon dots (N-CDs) display strong photosensitivity activity and increase the fluorescence features of the QDs to which they are conjugated, as shown in Fig. 3. N-CDs are also effective at increasing the fluorescence activity of photodynamic therapy against cancer cells [17]. In studies comparing QDs and the FITC-fluorescein isothiocyanate labelling technique to visualize Tca8113 tongue carcinoma cells, the QDs provided better fluorescence and photostability than the FITC labelling technique did, as shown in Fig. 4 [24].

4.2 *In Vivo Imaging Techniques*

QDs are used in cancer bioimaging because of their fluorescent signals and multiplex capabilities [27]. QDs are sensitive and selective in the detection of early-stage tumors. To understand the evolutionary mechanism of tumor progression in vivo, tumor imaging can be performed. N-CDs are also effective for the imaging of cancer, and in vivo studies have shown that N-CDs coupled with photodynamic therapy are very effective against oral cancer cells, as shown in Fig. 5 [28]. Compared with other traditionally used markers, such as fluorescent proteins and dyes, QDs are more stable, produce controllable photoluminescence, and can be easily modified in their chemical structure. Studies have been conducted for in vivo imaging of OSCC by using an EGFR monoclonal antibody conjugated with QDs in mice, as shown in

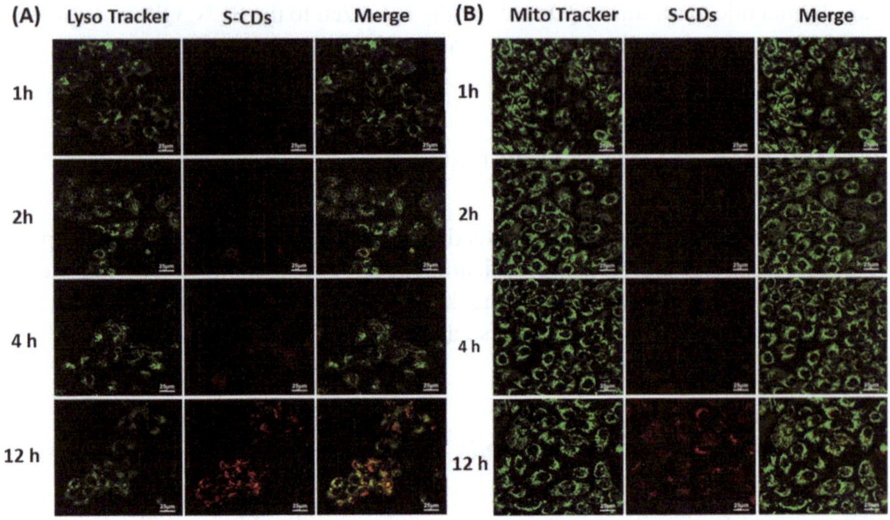

Fig. 2 Confocal fluorescence image of UM1 cancer cells treated with S-CDs. These images show that the S-CDs can be used as nanophotodynamic agents for oral cancer cells [26]

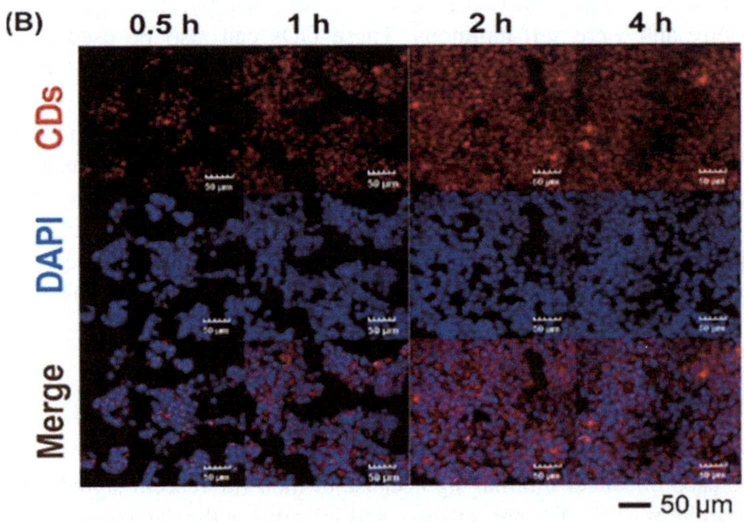

Fig. 3 The in vitro uptake of the carbon dots (CDs) by CAL-27 shows that the CDs are reliable fluorescent probes for bioimaging, as shown by the red fluorescence in the image [28]

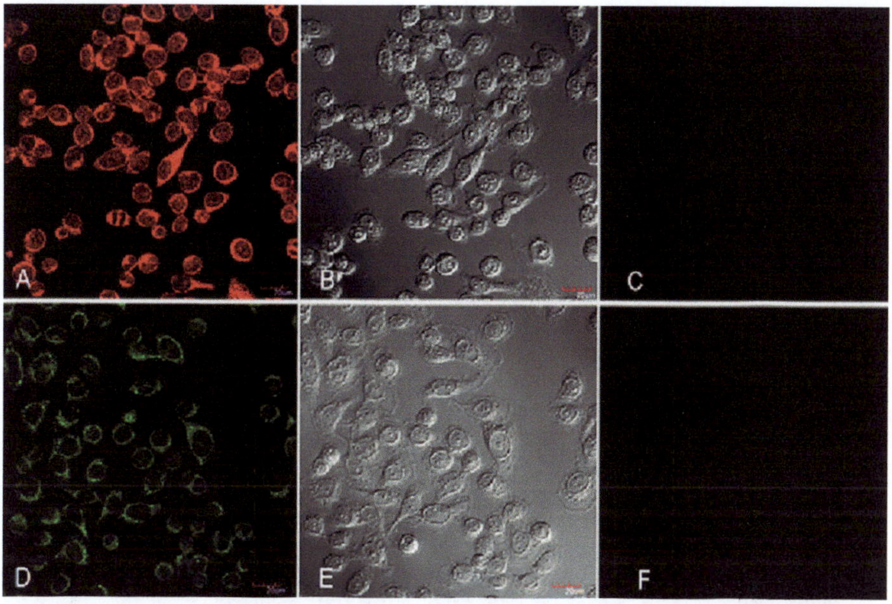

Fig. 4 One-color fluorescence imaging of HSP70 labelled with QD655 nm or FITC. (A) QD655 nm-tagged HSP70 and (D) FITC-tagged HSP70 exhibited identical lymphocyte placement, but the photofluorogram of the QD655 nm-tagged HSP70 had higher specificity and fluorescence intensity [24]

Fig. 6 [29]. Near-infrared luminescent QDs (NIR QDs) have been used in vivo for the visualization of OSCC BcaCD885 cells. The study revealed that the fluorescence intensity of the QDs for in vivo imaging increases as the number of QDs used increases, as shown in Fig. 7 [29]. Future studies can focus on the gradual decline in fluorescence intensity due to continuous cell division [29] (Table 2).

4.3 Use of QDs as Nanobiochips for the Detection of Oral Cancer

QDs are ideal for the analysis of oral samples because of the ability of nanobiochips to separate and measure biological analytes from sample matrices in small systems. To measure biological analytes, fluorescent transduction signals from nanoparticles are used [31–33]. Numerous studies have shown that QDs can be used to detect three important biomarkers, CA125, C-erbB2, and CEA, which are important for oral cancer detection [34–36].

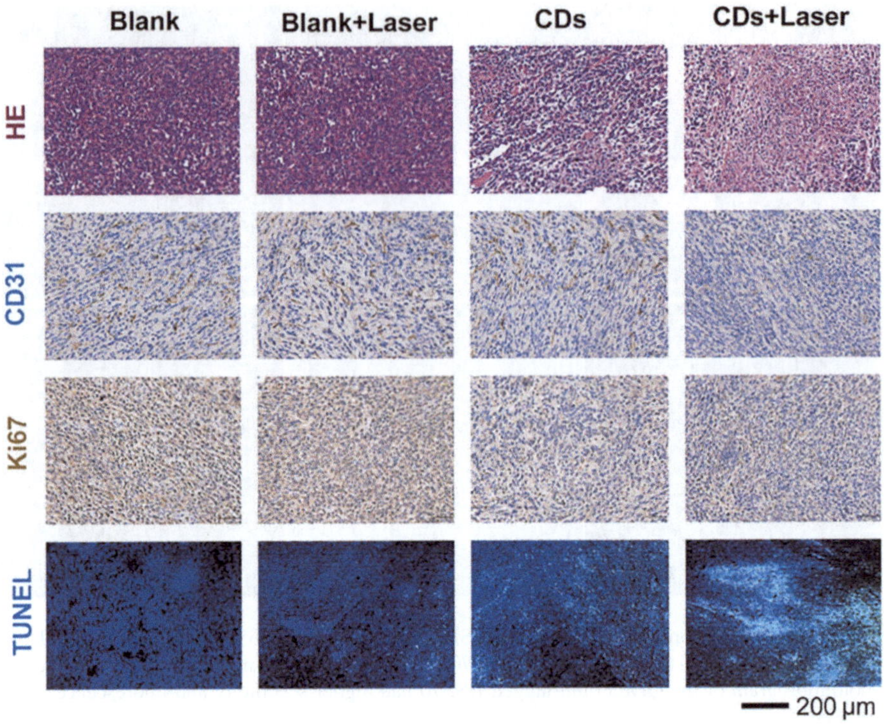

Fig. 5 Use of N-CDs for the in vivo imaging of CAL-27 cells in orthotropic tumor-bearing mice. The use of carbon dot-mediated photodynamic therapy was able to cause the apoptosis of OSCC cells [28]

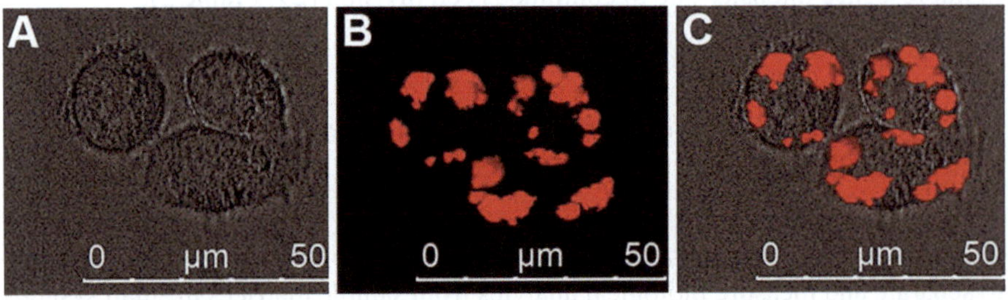

Fig. 6 In vivo labelling of BcaCD885 cells with the QD800-EGFR Ab probe. The use of QDs aids in the bioimaging of cancer cells [29]

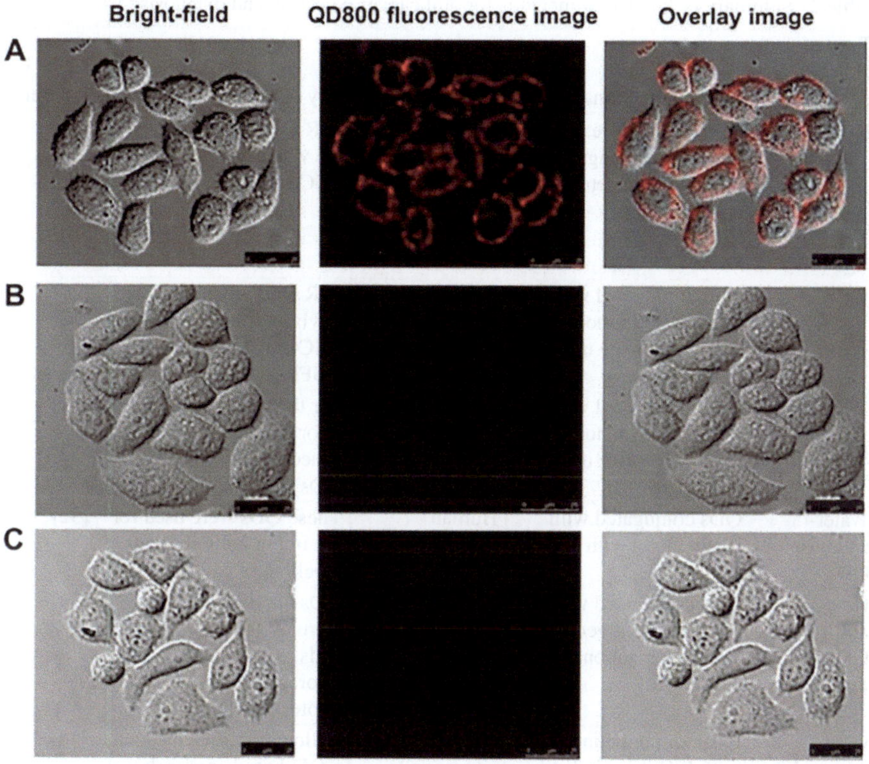

Fig. 7 In vivo imaging of OSCC using QDs. Fluorescence images of BcaCD885 cells after labelling with QDs-QD800. (A) Bright field image. (B) QD800 fluorescence image. (C) Overlay image [29]

4.4 Applications of QDs in Nanoenzymes

Nanoenzymes have been used for the early detection of the oral cancer microenvironment, tumor microenvironment, and tissue imaging [37]. Studies have been conducted in which individual encapsulated nanoparticles are packaged in phospholipid block copolymer micelles. These micelles have been used successfully for in vivo and in vitro imaging. Studies have shown that micelle-encapsulated QDs can be used for both in vivo and in vitro studies. QDs act as fluorescent semiconductor crystals that produce fluorescence and are stable in different types of biological environments [38].

Table 2 Summary of QD-based methods for oral cancer detection and diagnosis [30]

Type of quantum dot used	Surface functionalization	Cell line used	Key features	References
NIR QDs	These QDs were conjugated along with membrane-penetrating polypeptides.	Used in human cells-Human Oral squamous carcinoma BeaCD885 cells.	NIR QDs were used for the visualization of OSCC in vivo.	[29]
	QDs conjugated with anti-EGFR monoclonal antibodies were used.	Used in OSCC nude mice model.	NIR QDs were used for in vivo visible imaging of OSCC by targeting the EGFR.	[29]
	QDs conjugated with anti-EGFR monoclonal antibodies were used.	Orthotropic tongue cancer-bearing nude mice.	For the targeted imaging of orthotropic tongue cancer, Ag2Se–cetuximab QDs were used.	[31]
Water-soluble QDs	QDs conjugated with biotin along with PEG.	Human Tongue cancer cells Tca8113.	These QDs were used for the immunofluorescent labelling of cancer cells.	[32]
Goat anti-rabbit QD655 nm-IgG	QDs conjugated with HSP70 by antigen–antibody reaction.	Human Tongue cancer cells Tca8113.	QDs and FITC labelling were used in Tca8113 cells to compare the fluorescence intensity and photostability.	[24]
Goat anti-rabbit QD655 nm-IgG and goat anti-mouse QD655 nm-IgG	Quantum dot conjugated with IgG that binds to both HSF-1 and HSP70 by antigen–antibody reaction.	Human tongue cancer cells SC-25.	Following the heat shock, a QD-based approach was used for heat shock protein 70 and heat shock factor 1.	[30]
S-CDs	CQDs conjugated with sulfur were used.	UM1 cancer cells.	The study shows that S-CDs can be used as nanophotodynamic agents for oral cancer cells.	[26]
N-CDs	CQDs conjugated with nitrogen were used to increase the photosensitizing activity of the QDs and to increase the photodynamic therapy of oral cancer cells.	Human Oral mucosal keratinized epithelial cells (HoK)	The study shows that the fluorescent N-CDs showed high photosensitivity activity, and it improved the fluorescent properties of the QDs.	[28]

4.5 *Application of QDs as Nanosensors*

QDs are also used in the form of molecularly actuated plasmonic nanosensors that are made up of gold nanoparticles. These nanosensors have been used for sentinel node biopsy analysis after oral cancer resection. Studies have shown that anti-EGFR antibodies conjugated with gold nanoparticles have 600% greater affinity for the detection of oral malignant epithelial cells. This has been confirmed via the high optical signals in surface plasmon resonance scattering. This approach can aid in the early detection of oral cancer via saliva samples from oral cancer patients [30].

5 Challenges and Future Prospects

Some challenges associated with the use of QDs for cancer detection include the risk of cellular toxicity due to the generation of reactive oxygen species. QDs such as CdSe contain cadmium, which can be released into the body. There is also the risk of splitting the organic layer of the QDs, which could lead to the disclosure of the inorganic core. QDs are made from heavy metals. The release of metal ions such as cadmium ions can lead to issues. Most QDs have a polymer coating that makes them stable and stops the release of cadmium. There is a need for short-term and long-term toxicological clinical studies in animal models. Previous studies have shown that QDs do not affect cell growth under normal conditions [39]. Short-term studies have also shown that the administration of QDs does not affect the metabolism or behavior of animals [40]. Future studies can focus on how rapidly the QDs can be eliminated from the body, whether the QDs remain resident in the body, and in which tissues they reside. For the long-term safety of QDs, toxicological studies must be carried out. Safer alternatives that are free from cadmium and other toxic metals can also be developed [42]. Some examples of low-toxicity QDs involve the replacement of cadmium with zinc. These QDs are less sensitive to environmental changes, and they have a narrow emission spectrum that can cover most of the visible spectral window [25, 41]. There is a need for further optimization of the specificity and efficiency of the QDs used in cancer treatments. Importantly, QDs target only cancer cells and do not harm healthy cells. The complex nature of the cellular interactions in the tumor microenvironment also influences the progression rate of cancer and the therapeutic efficacy of QDs [42]. The cost of developing QD-based technologies is high, which can limit the accessibility and use of these techniques in widespread clinical practice [40]. Early and accurate detection of oral cancer can help save a patient's life. The correct diagnosis helps save time and can also reduce the negative impacts of invasive surgical interventions. Conventional methods such as surgical biopsy and histology are widely considered the gold standard, and new diagnostic methods such as QDs are the way forward. Newer screening methods can help detect cancer in its early stages. The development of QDs has provided effective cancer-imaging probes. QDs are inert and stable. The conjugation of QDs with biomolecules allows the active targeting of tumors.

Acknowledgement The authors of the present book chapter are thankful to the Manipal Academy of Higher Education, Manipal, India and Tezpur University, Assam, India for the research work.

References

1. Das, P., & Deshmukh, R. S. (2024). Quantum dots: The trailblazers of early detection. *Journal of Oral and Maxillofacial Pathology: JOMFP, 28*(1), 100–105. https://doi.org/10.4103/jomfp.jomfp_377_23
2. Tiwari, H., Rai, N., Singh, S., Gupta, P., Verma, A., Singh, A. K., Kajal, S. P., Singh, S. K., & Gautam, V. (2023). Recent advances in nanomaterials-based targeted drug delivery for preclinical cancer diagnosis and therapeutics. *Bioengineering (Basel, Switzerland), 10*(7), 760. https://doi.org/10.3390/bioengineering10070760
3. Smith, A. M., Dave, S., Nie, S., True, L., & Gao, X. (2006). Multicolor quantum dots for molecular diagnostics of cancer. *Expert Review of Molecular Diagnostics, 6*(2), 231–244. https://doi.org/10.1586/14737159.6.2.231
4. Abati, S., Bramati, C., Bondi, S., Lissoni, A., & Trimarchi, M. (2020). Oral cancer and precancer: A narrative review on the relevance of early diagnosis. *International Journal of Environmental Research and Public Health, 17*(24), 9160. https://doi.org/10.3390/ijerph17249160
5. Hamidu, A., Pitt, W. G., & Husseini, G. A. (2023). Recent breakthroughs in using quantum dots for cancer imaging and drug delivery purposes. *Nanomaterials, 13*(18), 2566. https://doi.org/10.3390/nano13182566
6. Montero, P. H., & Patel, S. G. (2015). Cancer of the oral cavity. *Surgical Oncology Clinics of North America, 24*(3), 491–508. https://doi.org/10.1016/j.soc.2015.03.006
7. Devi, S., Kumar, M., Tiwari, A., Tiwari, V., Kaushik, D., Verma, R., et al. (2022). Quantum dots: An emerging approach for cancer therapy. *Frontiers in Materials, 8*, 798440. https://doi.org/10.3389/fmats.2021.798440
8. Ghasemi, Y., Peymani, P., & Afifi, S. (2009). Quantum dot: magic nanoparticle for imaging, detection and targeting. *Acta biomedica : Atenei Parmensis, 80*(2), 156–165. PMID: 19848055.
9. Cai, W., Shin, D. W., Chen, K., Gheysens, O., Cao, Q., Wang, S. X., Gambhir, S. S., & Chen, X. (2006). Peptide-labelled near-infrared quantum dots for imaging tumor vasculature in living subjects. *Nano Letters*, *6*(4), 669–676. https://doi.org/10.1021/nl052405t.
10. Mousavi, S. M., Hashemi, S. A., Yari Kalashgrani, M., Omidifar, N., Lai, C. W., Vijayakameswara Rao, N., Gholami, A., & Chiang, W. H. (2022). The pivotal role of quantum dots-based biomarkers integrated with ultra-sensitive probes for multiplex detection of human viral infections. *Pharmaceuticals (Basel, Switzerland), 15*(7), 880. https://doi.org/10.3390/ph15070880
11. Chaturvedi, V. K., Singh, A., Singh, V. K., & Singh, M. P. (2019). Cancer nanotechnology: A new revolution for cancer diagnosis and therapy. *Current Drug Metabolism, 20*(6), 416–429. https://doi.org/10.2174/1389200219666180918111528
12. Sarkar, S., Srivastava, T. P., Sahoo, O. S., Shankar, A., Rai, A., Pethusamy, K., Dhar, R., & Karmakar, S. (2024). Applications of quantum dots in preventive oncology. *Asian Pacific Journal of Cancer Prevention : APJCP, 25*(3), 747–756. https://doi.org/10.31557/APJCP.2024.25.3.747
13. Singh, D., Thapa, S., Singh, K. R., Verma, R., Singh, R. P., & Singh, J. (2023). Cadmium selenide quantum dots and its biomedical applications. *Materials Letters: X, 18*, 100200. https://doi.org/10.1016/j.mlblux.2023.100200

14. Badıllı, U., Mollarasouli, F., Bakirhan, N. K., Ozkan, Y., & Ozkan, S. A. (2020). Role of quantum dots in pharmaceutical and biomedical analysis, and its application in drug delivery. *TrAC Trends in Analytical Chemistry, 131*, 116013. https://doi.org/10.1016/j.trac.2020.116013
15. Jana, P., & Dev, A. (2022). Carbon quantum dots: A promising nanocarrier for bioimaging and drug delivery in cancer. *Materials Today Communications, 32*, 104068. https://doi.org/10.1016/j.mtcomm.2022.104068
16. Helle, M., Cassette, E., Bezdetnaya, L., Pons, T., Leroux, A., Plénat, F., et al. (2012). Visualization of sentinel lymph node with indium-based near infrared emitting quantum dots in a murine metastatic breast cancer model, *7*. https://doi.org/10.1371/journal.pone.0044433
17. Yong, K. T., Ding, H., Roy, I., Law, W. C., Bergey, E. J., Maitra, A., & Prasad, P. N. (2009). Imaging pancreatic cancer using bioconjugated InP quantum dots. *ACS Nano, 3*(3), 502–510. https://doi.org/10.1021/nn8008933
18. Le, N., Zhang, M., & Kim, K. (2022). Quantum dots and their interaction with biological systems. *International Journal of Molecular Sciences, 23*(18), 10763. https://doi.org/10.3390/ijms231810763
19. Wagner, A. M., Knipe, J. M., Orive, G., & Peppas, N. A. (2019). Quantum dots in biomedical applications. *Acta Biomaterialia, 94*, 44–63. https://doi.org/10.1016/j.actbio.2019.05.022
20. Fauvel, B., & Yasri, A. (2014). Antibodies directed against receptor tyrosine kinases: Current and future strategies to fight cancer. *MAbs, 6*(4), 838–851. https://doi.org/10.4161/mabs.29089
21. Wu, X., Liu, H., Liu, J., Haley, K. N., Treadway, J. A., Larson, J. P., Ge, N., Peale, F., & Bruchez, M. P. (2003). Immunofluorescent labelling of cancer marker Her2 and other cellular targets with semiconductor quantum dots. *Nature Biotechnology, 21*(1), 41–46. https://doi.org/10.1038/nbt764
22. Rowinsky, E. K. (2004). The erbB family: targets for therapeutic development against cancer and therapeutic strategies using monoclonal antibodies and tyrosine kinase inhibitors. *Annual Review of Medicine, 55*, 433–457. https://doi.org/10.1146/annurev.med.55.091902.104433
23. Tan, W. B., Jiang, S., & Zhang, Y. (2007). Quantum-dot based nanoparticles for targeted silencing of HER2/neu gene via RNA interference. *Biomaterials, 28*(8), 1565–1571. https://doi.org/10.1016/j.biomaterials.2006.11.018
24. Zhao, J. J., Chen, J., Wang, Z. P., Pan, J., & Huang, Y. H. (2011). Double labelling and comparison of fluorescence intensity and photostability between quantum dots and FITC in oral tumors. *Molecular Medicine Reports, 4*(3), 425–429. https://doi.org/10.3892/mmr.2011.457
25. Fang, M., Peng, C. W., Pang, D. W., & Li, Y. (2012). Quantum dots for cancer research: current status, remaining issues, and future perspectives. *Cancer Biology & Medicine, 9*(3), 151–163. https://doi.org/10.7497/j.issn.2095-3941.2012.03.001
26. Li, Q., Zhou, R., Xie, Y., Li, Y., Chen, Y., & Cai, X. (2020). Sulfur-doped carbon dots as a highly efficient nanophotodynamic agent against oral squamous cell carcinoma. *Cell Proliferation, 53*(4), e12786. https://doi.org/10.1111/cpr.12786
27. Pandey, S., & Bodas, D. (2020). High-quality quantum dots for multiplexed bioimaging: A critical review. *Advances in Colloid and Interface Science, 278*, 102137. https://doi.org/10.1016/j.cis.2020.102137
28. Sun, L., Zhao, Y., Yan, J., Wang, J., Cheng, H., Wang, Y., et al. (2024). Nitrogen-doped carbon dots as a highly efficient photosensitizer for photodynamic therapy to promote apoptosis in oral squamous cell carcinoma. *Colloids and Surfaces A: Physicochemical and Engineering Aspects, 697*, 134409. https://doi.org/10.1016/j.colsurfa.2024.134409
29. Yang, K., Zhang, F. J., Tang, H., Zhao, C., Cao, Y. A., Lv, X. Q., et al. (2011). *In vivo* imaging of oral squamous cell carcinoma by EGFR monoclonal antibody conjugated near-infrared Quantum Dots in mice. *International Journal of Nanomedicine, 6*, 1739–1745. https://doi.org/10.2147/IJN.S23348
30. Chen, X. J., Zhang, X. Q., Liu, Q., Zhang, J., & Zhou, G. (2018). Nanotechnology: a promising method for oral cancer detection and diagnosis. *Journal of Nanobiotechnology, 16*(1), 52. https://doi.org/10.1186/s12951-018-0378-6

31. Zhu, C. N., Chen, G., Tian, Z. Q., Wang, W., Zhong, W. Q., Li, Z., Zhang, Z. L., & Pang, D. W. (2017). Near-infrared fluorescent Ag_2 Se-Cetuximab nanoprobes for targeted imaging and therapy of cancer. *Small (Weinheim a der Bergstrasse, Germany), 13*(3). https://doi.org/10.1002/smll.201602309
32. Li, Z., Wang, K., Tan, W., Li, J., Fu, Z., Ma, C., Li, H., He, X., & Liu, J. (2006). Immunofluorescent labelling of cancer cells with quantum dots synthesized in aqueous solution. *Analytical Biochemistry, 354*(2), 169–174. https://doi.org/10.1016/j.ab.2006.04.029
33. Soper, S. A., Brown, K., Ellington, A., Frazier, B., Garcia-Manero, G., Gau, V., Gutman, S. I., Hayes, D. F., Korte, B., Landers, J. L., Larson, D., Ligler, F., Majumdar, A., Mascini, M., Nolte, D., Rosenzweig, Z., Wang, J., & Wilson, D. (2006). Point-of-care biosensor systems for cancer diagnostics/prognostics. *Biosensors & Bioelectronics, 21*(10), 1932–1942. https://doi.org/10.1016/j.bios.2006.01.006
34. Rajguru, J. P., Mouneshkumar, C. D., Radhakrishnan, I. C., Negi, B. S., Maya, D., Hajibabaei, S., & Rana, V. (2020). Tumor markers in oral cancer: A review. *Journal of Family Medicine and Primary Care, 9*(2), 492–496. https://doi.org/10.4103/jfmpc.jfmpc_1036_19
35. Balan, J. J., Rao, R. S., Premalatha, B. R., & Patil, S. (2012). Analysis of tumor marker CA 125 in saliva of normal and oral squamous cell carcinoma patients: a comparative study. *The Journal of Contemporary Dental Practice, 13*(5), 671–675. https://doi.org/10.5005/jp-journals-10024-1207
36. Meng, Y., Yang, P., & Ma, L. (2020). Prognostic and clinical implications of c-erbB-2 expression in patients with oral cancer: A meta-analysis. *Medicine, 99*(23), e20575. https://doi.org/10.1097/MD.0000000000020575
37. Augustine, D., Sowmya, S. V., Gujjar, N., Pushpalatha, C., & Haragannavar, V. C. (2022). Role of nanozymes in oral cancer the road ahead. *Topics in Catalysis, 65*(19–20), 1973–1983. https://doi.org/10.1007/s11244-022-01729-9
38. Dubertret, B., Skourides, P., Norris, D. J., Noireaux, V., Brivanlou, A. H., & Libchaber, A. (2002). *In vivo* imaging of quantum dots encapsulated in phospholipid micelles , *298*(5599), 1759–1762. https://doi.org/10.1126/science.1077194
39. Derfus, A. M., Chan, W. C. W., & Bhatia, S. N. (2004). Probing the Cytotoxicity Of Semiconductor Quantum Dots. *Nano Letters, 4*(1), 11–18. https://doi.org/10.1021/nl0347334
40. Ballou, B., Lagerholm, B. C., Ernst, L. A., Bruchez, M. P., & Waggoner, A. S. (2004). Noninvasive imaging of Quantum Dots in mice. *Bioconjugate Chemistry, 15*(1), 79–86. https://doi.org/10.1021/bc034153y
41. Pradhan, N., Battaglia, D. M., Liu, Y., & Peng, X. (2007). Efficient, stable, small, and water-soluble doped ZnSe nanocrystal emitters as noncadmium biomedical labels. *Nano Letters, 7*(2), 312–317. https://doi.org/10.1021/nl062336y
42. Omidian, H., Wilson, R. L., & Cubeddu, L. X. (2024). Quantum Dot Research in Breast Cancer: Challenges and Prospects. *Materials (Basel, Switzerland), 17*(9), 2152. https://doi.org/10.3390/ma17092152

Quantum Dots for Photodynamic Therapy

Sachin Shetty, Ayush Khandelwal, Aishath Shaheeda, Mrudula Gore, Roopal Behera, and Shama Prasada Kabekkodu

1 Introduction

Quantum dots (QDs) are nanoscale crystals that are fluorescent semiconductor colloids that were synthesized for the first time in the early 1980s [1]. These artificial semiconductor nanoparticles generally possess characteristic optical, electronic, and photophysical properties that render them suitable for a number of potential applications in fluorescent biological tagging, imaging, photovoltaic cells, hybrid materials, and sensing in addition to being optimal fluorescence resonance energy transfer (FRET) donor agents [2]. Semiconductor particles small enough exhibit quantum confinement effects, which determine the energies that are available in the particles for the formation of electron–hole pairs. The energy and wavelength of light, which correspond to specific colors, are directly linked, allowing the optical properties to be finely tuned by altering the particle size. By controlling the size of QDs, particles can be designed to emit or absorb light at specific wavelengths, producing a variety of colors [3].

The Latin term "quantum" refers to the smallest quantity of any physical entity participating in an interaction. In the field of quantum physics, QDs exhibit different optical and electrical characteristics than larger particles do. It is possible to excite an electron in a QD to a higher energy state by exposing it to UV light. When an excited electron returns to the valence band, it emits light (luminescence) and releases its energy. The color of this light is determined by the energy gap between the valence and conduction bands [4]. QDs were first discovered in the late 1970s by the Russian physicist Alexei Ekimov, who synthesized copper chloride and

S. Shetty · A. Khandelwal · A. Shaheeda · M. Gore · R. Behera · S. P. Kabekkodu (✉)
Department of Cell and Molecular Biology, Manipal School of Life Sciences, Manipal Academy of Higher Education, Manipal, Karnataka, India
e-mail: mrudula.mslsmpl2023@learner.manipal.edu; roopal.mslsmpl2023@learner.manipal.edu; shama.prasada@manipal.edu

S. P. Kabekkodu et al. (eds.), *Functional Quantum Dots*, Lecture Notes in Nanoscale Science and Technology 40,
https://doi.org/10.1007/978-3-031-98219-4_8

subsequently cadmium selenide nanocrystals within a molten glass matrix. He observed fluorescence and color changes, publishing his findings in 1980 [5]. In 1982, another Russian physicist, Alexander Efros, proposed a theory explaining the behavior of these nanoscale crystals through electron confinement. Inspired by Ekimov's work, Bell Labs successfully created nanocrystals in liquid form, producing a colloidal suspension. By 1983, they synthesized colloidal QDs of cadmium sulfide, which are more practical to handle, and documented their results [6]. Because of their many potential uses and strong economic appeal, QDs represent a rapidly expanding field of study. QDs exhibit quantum confinement when their particle size is smaller than the Bohr radius of the electron–hole pair (excitons). In this confined state, the energy levels become quantized, akin to the "particle in a box" model. The discrete energy levels in such QDs are size dependent, directly influencing their bandgap energy [7]. Additionally, when the size of the QDs is smaller than the de Broglie wavelength, a correlation emerges between their optical and electrical properties. As highly luminescent materials, QDs demonstrate a broad excitation range and narrow, symmetric emission spectra after excitation, which are significantly sharper than those of conventional fluorophores or organic dyes. Moreover, when light with energy matching the bandgap is absorbed, it excites an electron, transitioning it from the valence band to the conduction band. The exited electron subsequently recombines and drops straight to the valence band, and in this process, it releases a photon [8].

QDs typically consist of three main components: a core, a shell, and a surface coating. These components work together to increase luminescence, improve stability, and activate the surface. Structurally, most QDs are of the core–shell variety. The core, which forms the foundation, is made up of monolayers of semiconductor material, often derived from an organic molecular framework such as CdSe or CdTe [9]. Quantum confinement phenomena in QDs are based on quantum mechanics. Understanding the wave–particle duality of matter and energy quantization in constrained systems is necessary to comprehend the electron behavior in QDs. In bulk semiconductor materials, electrons move freely inside a three-dimensional lattice structure. Any energy level can be occupied by electrons in these continuous materials. Nevertheless, electron mobility is constrained in all three dimensions by nanoscale semiconductor materials [10]. QDs have special properties not present in bulk semiconductors because of their quantized energy levels. The emission spectra of QDs are sharp and narrow because they emit light at certain wavelengths on the basis of their size and structure. Because of their emission spectrum tenability, QDs have potential use in solar cells and LEDs [11].

2 Quantum Dot Synthesis

QDs can be synthesized through various techniques, each offering distinct benefits and limitations. Some of the most commonly employed methods include the following:

2.1 Colloidal Synthesis

It is a widely used method for producing QDs and involves the nucleation and growth of nanocrystals in a solvent. This technique allows precise control over the size and composition of QDs by adjusting reaction parameters such as temperature, precursor concentration, and reaction time. The process typically includes the following steps: a.) nucleation (rapid formation of small clusters from precursors), b.) growth (controlled enlargement of these clusters into nanocrystals), and c.) surface passivation (stability of QDs by capping agents to prevent aggregation and improve their solubility). This method is versatile and can produce QDs with high crystallinity and tunable optical properties [12].

2.2 Microfluidic Synthesis

This employs microscale fluid channels to produce QDs, offering enhanced control over the reaction conditions compared with traditional batch synthesis. The advantages of this method include the following:

(a) Precise Control: Accurate manipulation of reaction parameters such as temperature and mixing rates,
(b) Uniformity: Production of QDs of consistent size and shape,
(c) Scalability: Potential for continuous production, facilitating large-scale manufacturing [13].

2.3 Biosynthesis

Biosynthesis involves the use of biological organisms or biomolecules to produce QDs, offering an eco-friendly alternative to conventional chemical methods. Various species, including bacteria, fungi, and plants, have been utilized for the biosynthesis of different types of QDs. This method provides a.) eco-friendliness: reduction in the use of toxic chemicals and b.) biocompatibility: potential for medical and biological applications due to reduced toxicity [14].

2.4 Viral Assembly

This method utilizes genetically engineered viruses to organize inorganic nanocrystals into QDs. This method leverages the specific recognition properties of viruses to create ordered arrays of QDs, allowing for a.) specificity: precise assembly of QDs at the nanoscale, and b.) functionalization: potential for creating complex nanostructures with specific properties [15].

Table 1 Types of QDs used in the biomedical field

S.No.	Types of quantum dot	Examples	Biomedical applications	References
1.	Cadmium-based QDs	CdSe, CdTe, CdS	Bioimaging: Used for labelling and imaging of cells and tissues due to their bright fluorescence. Drug delivery: Serve as carriers for targeted drug delivery systems. Theranostics: Combine therapeutic and diagnostic functions.	[17, 18]
2.	Carbon-based QDs	Graphene quantum dots, carbon Nanodots	Bioimaging: Offer biocompatibility and low toxicity for imaging applications. Drug delivery: Potential use in delivering drugs due to their functionalizable surfaces.	[19]
3.	Silicon QDs	SiQDs	Fluorescence imaging: Utilized for in vitro and in vivo imaging owing to their biocompatibility and near-infrared emission. Time-gated imaging: Employed in techniques that require long luminescent lifetimes.	[20]
4.	Perovskite QDs	$CsPbBr_3$, $MAPbI_3$	Bioimaging: Emerging use in imaging due to their high brightness and tunable emission wavelengths.	[21]
5.	Indium-based QDs	InP, InAs	Bioimaging: Serve as less toxic alternatives to cadmium-based QDs for imaging applications.	[22]

2.5 Electrochemical Assembly

This approach is employed to produce highly organized QD arrays. The process begins with an ionic reaction at the electrolyte–metal interface, which facilitates the spontaneous assembly of nanostructures on the metal surface. The key advantages of this method include a) precision, i.e., the creation of well-ordered QD arrays, and b) scalability, i.e., the ability for large-scale production [16]. Table 1 presents various types of QDs.

3 Quantum Dot Biosynthesis: Top-Down and Bottom-up Approaches

QDs can be synthesized via physical, chemical, or biological methods. However, the biological synthesis method is more efficient than the physical and chemical approaches are because of its surface defects and stringent and toxic reaction

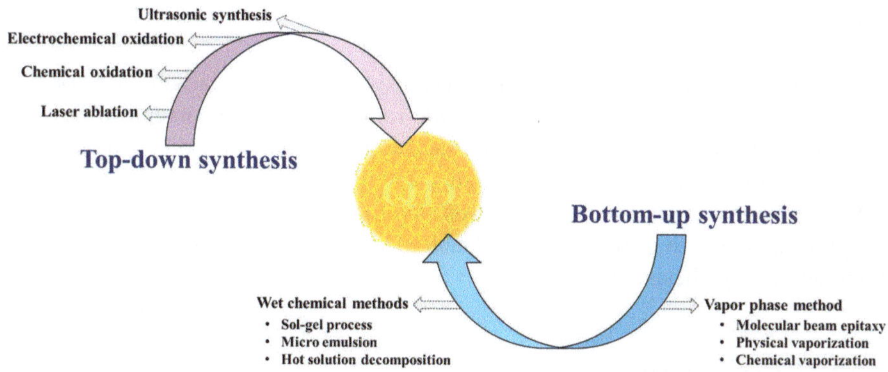

Fig. 1 Diagrammatic representation of the top-down and bottom-up methods of quantum dot synthesis

conditions [23]. Generally, top-down or bottom-up techniques are used for the synthesis of QDs. The top-down approach involves yielding nanoparticles from bulky substrates, whereas bottom-up approaches involve the use of molecular precursors for the assembly of QDs (Fig. 1).

3.1 Top-Down Synthesis

In this approach, QDs of approximately 30 nm are synthesized from bulky semiconductors [24]. Top-down processing methods include electrochemical oxidation, chemical oxidation, laser ablation, and ultrasonic synthesis [25].

3.1.1 Electrochemical Oxidation

Electrochemical oxidation is a cost-efficient method of QD synthesis that offers high purity as well as high yield along with scalability and reproducibility. Here, the graphite rods or carbon nanotubes are subjected to electrochemical cutting via an oxidation process. The electrolytes and applied current determine the surface properties of the QDs, and the oxygenation process determines the optical characteristics. Even though the operation process is complicated, this method offers adaptability on the basis of current and pH, which can be exploited for application-specific QDs [25].

3.1.2 Chemical Oxidation

Chemical oxidation results in the bulky synthesis of QDs from carbon-based precursors via strong oxidation reactions. A wide range of carbon sources can be used, such as coal-activated carbons, coconut-activated carbons, or wood-activated carbons, and are etched via strong oxidizing agents such as nitric acid. This step is followed by an amine-based passivization step for enhancing the optical and surface properties. Owing to its high yield and lack of sophisticated devices, it is a versatile approach for synthesis [26].

3.1.3 Laser Ablation

Laser ablation uses high-energy laser beams to break a target material such as graphite to produce QDs. It is a versatile technique for synthesis because of its easy morphology control and tunability. High energy from the laser in controlled air and vapor conditions is used to heat the target material to form nanocrystals. In liquid media, a high-energy laser pulse results in the formation and expansion of bubbles. As the laser pulse ends, the bubbles shrink to form a cluster of carbon, resulting in the formation of QDs [27].

3.1.4 Ultrasonic Synthesis

Ultrasound synthesis is an eco-friendly method for the synthesis of QDs under mild experimental conditions. Here, ultrasonic waves are used for the generation of extreme conditions to fragment bulky graphites or semiconductors. High-frequency waves are used to generate a cycle of high- and low-pressure regions in the medium, which results in the formation and collapse of cavitation bubbles. As these bubbles carry very high energy, their collapse results in breaking down the substrate to form QDs. Even though this method offers scalable and tunable manufacturing of QDs under green conditions, sophisticated equipment is needed for synthesis.

3.2 Bottom-Up Synthesis

Bottom synthesis uses a self-assembly technique to synthesize precise QDs with limited defects. Moreover, bottom-up synthesis is a complex process, as it utilizes complex organic solvents to construct nanoparticles. The process can be broadly divided into wet chemical methods and vapor phase techniques [28].

3.2.1 Wet Chemical Methods

The wet chemical method involves a nucleation phase and growth phase of nanoparticles via conventional precipitation methods. The common synthesis process includes techniques such as the sol–gel process, microemulsion, and hot solution decomposition.

The sol–gel process involves the synthesis of nanoparticles from metal precursors such as alkoxides, acetates, or nitrates. Upon hydrolysis, these metal precursors generate nanoparticles in a medium called sol, which then condense to form a gel [24]. Even though it is a simple process, it is subjected to high-concentration defects, limiting its application.

The microemulsion technique involves the formation of a microemulsion by mixing two immiscible solutions that are stabilized by surfactants. The molar ratio of the liquid to the surfactant determines the growth and shape of the QDs; hence, the size can be controlled by changing the molar ratio. However, the yield might be lower, and there is a high chance of impurities and defects in the synthesized nanoparticles [29].

In hot solution decomposition, high-temperature pyrolysis of organometallic substrates in a coordinating solvent is used for the synthesis of QDs. This process involves the preparation of a coordinating solvent and rapid injection of metal precursors into it for nucleation, followed by growth and stabilization of the particles. This method provides high-quality QDs, as high thermal energy minimizes defects through thermal annealing [30].

Overall, the wet chemical method is a versatile and scalable method for quantum dot synthesis; however, the use of potentially toxic compounds is considered a limitation.

3.2.2 Vapor Phase Method

The vapor phase method is another sophisticated bottom-up approach in which atoms are deposited atom by atom to form QDs. The growth of the QDs over a substrate then proceeds in layer-by-layer form, island forms, or a combination of both. This process can involve either molecular beam epitaxy or physical or chemical vapor deposition [31].

In molecular beam epitaxy, substrate elements such as arsenic and gallium are vaporized, and their condensation under controlled conditions favors the formation of QDs. Physical vaporization involves the melting of a solid to a liquid and its vaporization to a low-pressure gas. For evaporation, electron beam heating, laser or arc discharge is used. In the chemical vaporization method, the precursors and heated substrates are subjected to a chemical reaction under controlled conditions to form a solid film deposition through surface energy dynamics. The vapor phase method is efficient in terms of manufacturing without prepatterning, but the inconsistent quantum dot size might result in altered optoelectronic properties.

4 Photophysical and Photochemical Properties of Quantum Dots

4.1 Photophysical Properties

Luminescence lifespan, luminescence yield, fluorescence and phosphorescence spectra, and electronic absorption spectra are typical photophysical characteristics.

4.1.1 Quantum Confinement Effect

One essential characteristic of a material is the band gap between its valence and conduction bands. Because of quantum confinement, the energy gap in QDs is size dependent and widens as the nanoparticle size decreases. The bandgap widens as QDs decrease below a threshold, which causes their emission and absorption to shift to higher energies (blueshift). Unlike the continuous energy bands found in bulk semiconductors, this phenomenon results in discrete energy levels in QDs (Fig. 2) [32].

4.1.2 Excitons and Quantum Confinement

An exciton (an electron–hole pair) is created when an electron is excited from the valence to conduction band by photons with energy greater than or equal to the bandgap energy. Because of the particle size, the electrons and holes are spatially constrained in QDs, making this effect especially apparent. As matter shrinks to the nanoscale, especially in semiconductors, quantum size effects (QSEs) occur. These impacts result in special physicochemical characteristics that distinguish them from

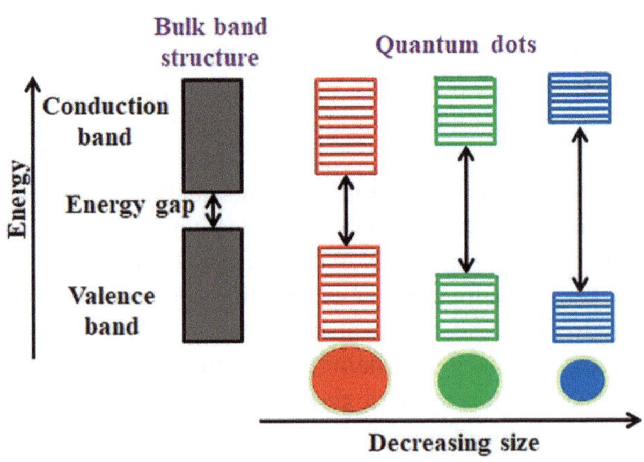

Fig. 2 Varying the band gap in relation to the size of the quantum dots

bulk materials. The size of the QD and how it interacts with its surroundings determine how these charge carriers behave inside QDs [33].

4.1.3 Tunability of the Band Gap

Band gap tunability is one of the distinctive optical characteristics of QDs and has received much scientific attention. Quantum confinement is the reason for tunability. The energy levels separate into distinct states as the nanoparticle size decreases to below the Bohr exciton radius, which is normally between 1 and 10 nm. The size-tunable absorption spectra and luminescence colors of QDs are caused by the discrete character of the energy levels in the nanostructures [32].

4.1.4 Fluorescence Emission and Absorption

The size of QDs has a significant effect on their optical characteristics. In particular, the Brus model explains that smaller QDs absorb and emit light at shorter (blue) wavelengths, whereas larger QDs cause the absorption onset and fluorescence emission wavelengths to increase (red) wavelengths. Because of this occurrence, QDs can be used in a variety of applications where exact control over the emission wavelength is crucial, such as fluorescent labelling and imaging.

- **Impact on CT and FRET:** By varying the QDs' size, one can modify the energy levels to promote charge transfer to or from redox-active complexes (for CT) or regulate the overlap of emission spectra with acceptor molecule absorption spectra (for FRET).
- **FRET Efficiency:** The energy transfer efficiency increases with increasing proximity of the QD's emission to the absorption maximum of the acceptor molecule. The emission wavelength of a QD can be adjusted to match a certain donor–acceptor pair perfectly [34].

4.1.5 Fluorescence Lifetime

Another significant optical characteristic of QDs that is influenced by their size is their fluorescence lifespan. The energy levels of larger QDs are more closely spaced, and the electron–hole pair has a greater chance of remaining in an excited state for a longer amount of time before relaxing to the ground state and releasing light. This increases the fluorescence lifetime.

- **Smaller QDs:** These dots have more widely spaced energy levels, which results in faster relaxation times and shorter fluorescence lifetimes. Because of their size-dependent lifetime, QDs are useful for applications in fluorescence-based sensing and imaging.

- **Larger QDs:** These dots have a longer fluorescence lifetime because of the longer-lived excitons made possible by the increased separation of energy levels in large dots [35].

4.1.6 Core–Shell QDs

It is possible to engineer QDs with core–shell architectures to increase the fluorescence quantum yield, which measures the effectiveness of light emission from an excited state. These structures consist of a semiconductor material with a wider band gap enclosing a core material with a smaller band gap. This method can improve the optical characteristics of QDs in several ways:

- **Reduced Surface Recombination:** By passivating the quantum dot surface, the shell can prevent electron–hole pairs from recombining nonradiatively, which could otherwise lower the quantum yield.
- **Auger Recombination Reduction:** The shell can also aid in reducing Auger recombination, which has a detrimental effect on the quantum yield and occurs when energy is transferred to a third charge carrier rather than being released as light [32].

The stability and fluorescence emission of QDs are greatly improved by this core–shell structure, increasing their effectiveness in solar cells, LEDs, and bioimaging.

4.2 Photochemical Properties

4.2.1 Surface Reactivity

The surface of quantum dots is essential for regulating their reactivity, stability, and optical characteristics. The surface is usually stabilized by surfactants or capping ligands, which are divided into three categories according to their electronic interactions [36]:

- **L-type ligands:** Neutral electron pair donors (such as amines and phosphines).
- **X-type ligands:** Monovalent anions (such as carboxylates and thiolates).
- **Z-type ligands:** Lewis acids that interact with electron-rich sites.

The choice of capping ligand affects the emission properties; for example, primary amines with long aliphatic chains can enhance the emission intensity in comparison to sterically hindered ligands, and smaller ligands, such as chloride anions, have been demonstrated to effectively passivate surface trap states.

4.2.2 Ligand Chemistry

A further improvement in the quenching efficiency for both FRET and CT interactions is the development of various ligands, such as poly(ethylene glycol) (PEG) chains with varying molecular weights, which provide fine control over the distance between the QD and the acceptor molecules [32].

4.2.3 High Surface-to-Volume Ratio

QDs can bind several molecules simultaneously because of their enormous surface area.

- **Multichannel Interactions:** Since the total rate of energy transfer increases with increasing number of acceptor molecules attached to a single QD, the energy transfer efficiency greatly increases. Likewise, for CT, the rate of charge transfer is increased by a greater density of redox complexes surrounding the QD.
- **Effect on FRET and CT:** The distance between the donor and acceptor molecules determines the degree to which FRET and CT work. These interactions may occur more quickly in smaller QDs because of their shorter center-to-center separation. Additionally, the size of the capping ligands or linkers that join the QD to the target molecules can be changed to adjust the distance. The interaction strength may be precisely controlled owing to this fine-tuning [32, 35].

4.2.4 Passivation

Because organic ligands are usually used to passivate QDs, the surface-to-volume ratio is high. This might result in the creation of trap states, which lowers the efficiency of the QD. QDs are frequently built with core–shell architectures to increase photostability and address issues such as surface trap states. Surface passivation with a larger bandgap semiconductor shell (e.g., ZnS for CdSe) enhances the stability and photoluminescence quantum yield [33].

- **Type I Structure:** Because of the shell's larger band gap, the electron–hole pair is contained within the core of core–shell QDs such as CdSe/ZnS.
- **Type II Structure:** The electron and hole are located in the core and shell, respectively, and are spatially separated in materials such as ZnTe/CdSe or CdS/ZnSe. Charge separation is encouraged, which is advantageous for processes such as photocatalysis.
- **Inverse Type I Structure:** Charge carriers are confined within the shell of materials such as ZnSe/CdSe or CdS/HgS, and the thickness of the shell controls the optical characteristics, such as the wavelengths of absorption and emission.

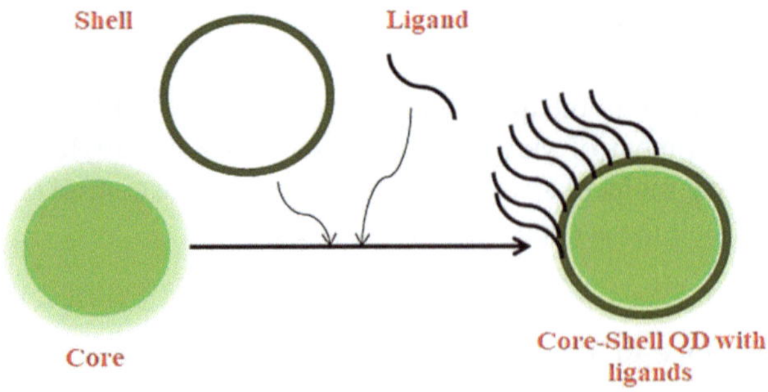

Fig. 3 Core-Shell architecture

Furthermore, surface ligands are necessary for QD dispersion and colloidal stability. For instance:

- **Hydrophobic ligands:** For QD production, organic solvents such as oleylamine or oleic acid are used.
- **Hydrophilic ligands:** In aqueous solutions, glutathione (GSH), thioglycolic acid (TGA), and mercaptopropionic acid (MPA) are utilized [33] (Fig. 3).

4.2.5 Surface Engineering for Water Solubility

Hydrophobic QDs are created by a variety of synthetic techniques, and they dissolve in nonpolar organic solvents such as chloroform or toluene. Surface changes such as ligand exchange are employed to make QDs hydrophilic. For example, species based on lipoic acid have shown promise in producing water-soluble QDs [32].

4.2.6 Self-Assembly and Functionalization

Through interactions such as hydrogen bonds or van der Waals forces, molecules can self-assemble onto QDs as part of another tactic. This may make it possible to functionalize QDs for particular uses, such as biological or sensing applications [36].

5 Applications in QDs

QDs are widely recognized for their versatility and functionality in various biological contexts, including but not limited to bioimaging and diagnostics, in vivo imaging, drug delivery, photodynamic therapy (PDT), biosensors, light indicators for biological coding, and various other applications.

5.1 In Vitro Applications of QDs

QDs exhibit remarkable potential in in vitro applications, leveraging their unique optical and electronic properties to advance diagnostic, bioimaging, and therapeutic strategies. The exceptional brightness, stability, and tunable fluorescence of these materials make them indispensable tools for precision biological research.

5.1.1 Diagnostics

QDs play a transformative role in early disease detection through highly sensitive diagnostic methodologies. Compared with traditional fluorescent dyes, functionalized QDs conjugated with targeting ligands can specifically bind to biomarkers, such as cancer markers, on cell surfaces, offering superior sensitivity. Their tunable emission properties enable multiplexed detection of multiple biomarkers within a single assay. Zhang et al. [37] developed an advanced QD-based immunosensor for detecting C-reactive protein (CRP), a critical biomarker for inflammation. This novel system utilizes a magnetic-driven, ratiometric fluorescence mechanism. By integrating fluorescent nanobeads with magnetic properties, the sensor achieved enhanced sensitivity and reduced detection time, making it highly suitable for rapid, point-of-care diagnostics [37].

5.1.2 Fluorescent Labelling in Bioimaging

QDs are widely used in cellular imaging applications because of their bright fluorescence and resistance to photobleaching, enabling prolonged and precise imaging. Kajani et al. [38] synthesized heteroatom-doped magneto-fluorescent carbon dots that facilitated the imaging of human fibroblasts in vitro, demonstrating rapid cellular uptake and superior imaging resolution [39]. Praseetha et al. [40] presented an innovative approach by creating supraneoparticle assemblies using fluorescent polymers and QDs for targeted immunolabelling, increasing imaging precision [39]. Praseetha et al. [40] demonstrated green-synthesized carbon QDs with excellent biocompatibility and fluorescence properties, enabling advanced theranostic imaging of 3 T3-L1 adipocytes [40].

5.1.3 Drug Delivery Systems

QDs have emerged as effective nanocarriers for therapeutic agents, enabling controlled and targeted drug delivery with simultaneous imaging capabilities (theranostics). Functionalized QDs deliver drugs to specific tissues or diseased cells, minimizing off-target effects. Upadhyaya et al. (2024) [41] synthesized nitrogen-doped carbon quantum dots (NCQDs) for curcumin delivery. These NCQDs improved the stability and cellular uptake of curcumin in live cell models, demonstrating their potential in therapeutic applications [42].

5.1.4 Cytotoxicity and Biocompatibility Studies

Evaluating the biocompatibility and cytotoxicity of QDs is critical for their safe application in biological systems. Khobrekar et al. [43] examined citric acid-functionalized neomycin carbon dots, confirming their minimal cytotoxicity and suitability for in vitro cellular applications [43].

5.1.5 Antioxidant and Antimicrobial Applications

QDs synthesized through green methods exhibit promising antioxidant and antimicrobial properties. Kadyan et al. [44] developed graphene quantum dots (GQDs) with enhanced synergistic antioxidant and antibacterial effects [44]. Khan et al. [45] applied carbon dots for microbial cell staining, demonstrating their efficacy in bacterial bioimaging and fluorescence analysis [45].

5.1.6 In Vitro Glucose Monitoring

QDs are utilized in biosensing platforms for precise and sensitive glucose monitoring. Kumar et al. [46] designed carbon QDs encapsulated in cellulose derivatives for fluorescence-based glucose detection. Their system achieved a detection limit as low as 7 nM, highlighting its potential in real-time glucose monitoring.

5.1.7 Photothermal Therapy

QDs are increasingly explored for photothermal therapeutic applications because of their capacity to convert light energy into localized heat. Cho et al. [47] synthesized glucose-derived carbon dots suitable for both imaging and photothermal therapy, which demonstrated significant efficacy in treating prostate cancer cells in vitro [47].

5.1.8 Advanced Cellular Probes

QDs enable advanced cellular imaging and analytical applications through their superior optical properties. Ramajayam et al. [48] synthesized boron- and nitrogen-doped carbon dots for the fluorimetric detection of Au^{3+} ions. These QDs exhibited high biocompatibility and were successfully employed for precise cell imaging [49].

5.2 *In Vivo Applications of QDs*

5.2.1 Diagnostics

QDs have demonstrated remarkable utility in enhancing diagnostic precision in vivo because of their superior optical properties.

5.2.1.1 Cancer Detection

The literature has shown that QDs are used for the detection of cancer. A study conducted by Li et al. [50] highlighted the integration of QDs with imaging technologies for effective tumor localization, leveraging their strong fluorescence even in deep-tissue environments [51]. Gadolinium-doped graphene QDs (Gd-GQDs) have been identified as multifunctional agents for both diagnostic imaging and therapeutic applications in oropharyngeal cancer [52]. Cadmium-free QDs emitting near-infrared fluorescence were shown to offer reduced toxicity and improved imaging penetration depth, enhancing tumor detection capabilities [53].

5.2.1.2 Liver Enzyme Activity Monitoring

A study conducted by Liu et al. [54] synthesized nitrogen-doped graphene quantum dots (N-GQDs), enabling early-stage liver disease detection via fluorescence-based sensors [54].

5.2.1.3 Guided Surgery

Jena et al. [55] developed enzyme-responsive QDs for surgical guidance, allowing precise intraoperative diagnostics [55].

5.2.2 In Vivo Bioimaging and Visualization

The exceptional fluorescence and stability of QDs have facilitated advanced imaging techniques in live animal models.

5.2.2.1 Enhanced Tumor Imaging

Zhang et al. [56] introduced red-emitting carbon dots with superior penetration and imaging clarity [57]. Shu et al. [58] demonstrated the utility of ytterbium- and gadolinium-doped carbon dots for multimodal imaging, combining fluorescence, magnetic resonance, and computed tomography for high-resolution tumor visualization [58].

5.2.2.2 Gastric Cancer Imaging

A probe based on CsPbBr3 perovskite QDs conjugated with specific peptides was developed for targeted imaging of gastric cancer, providing exceptional fluorescence and precision [37].

5.2.2.3 Deep-Tissue Imaging

Matea et al. [59] reviewed the application of QDs as fluorescent probes, emphasizing their superior signal–to–noise ratios in deep-tissue imaging [59].

5.2.3 Cancer Therapy and Drug Delivery

Owing to their tunable photophysical properties and biocompatibility, QDs have been employed as versatile platforms for cancer therapy and targeted drug delivery.

5.2.3.1 Drug Delivery Systems

Dar and Sahu [60] utilized graphene quantum dots (GQDs) to increase the solubility and control the release of curcumin, resulting in significant tumor inhibition [60]. Song et al. [61] reported enzymatic cascade nanoreactors based on carbon dots for tumor targeting, enabling self-enhancing drug delivery and synergistic therapy [62].

5.2.3.2 Photothermal and Photodynamic Therapy

Reduced graphene oxide quantum dots (RGQDs) loaded with immunomodulatory agents demonstrated effective photothermal therapy in glioblastoma by enhancing the immune response and inducing tumor cell destruction [63]. Araújo-Silva et al. [64] synthesized nanographene oxide conjugates for photodynamic therapy, achieving high drug-loading efficiency and precise tumor targeting [65].

5.2.3.3 Reactive Oxygen Species (ROS)-Mediated Therapy

Wang et al. [66] developed carbon dots conjugated with cisplatin (HA-CDs-DDP) to induce apoptosis in oral tongue squamous cell carcinoma through ROS generation [66]

5.2.3.4 Bone Regeneration

Yan et al. [67] combined graphene oxide QDs with dental stem cells, facilitating mitophagy-regulated bone regeneration in vivo [67].

5.2.4 Neuronal Imaging and Cellular Analysis

QDs are advancing the study of neuronal systems and cellular dynamics. Kanemaru et al. [68] utilized fluorescent QD biosensors to monitor dynamic cellular signalling, enabling high-resolution tracking of insulin granules in vivo [68].

5.2.5 Toxicity and Biocompatibility Evaluation

Evaluating and mitigating the toxicity of QDs remains a critical focus for their broader biomedical applications [69].

5.2.5.1 Nephrotoxicity and Surface Chemistry

Jiang et al. [70] examined the nephrotoxicity of carbon QDs and the role of surface charge in biocompatibility, providing insights for safer designs [70].

5.2.5.2 Nontoxic Probes

Kumar et al. [46] demonstrated the nontoxic behavior of biomass-derived carbon dots as quantum probes for visualizing inflammation in vivo [46].

5.3 Other Applications

QDs possess remarkable optical and electronic properties, making them invaluable across diverse biological applications that extend beyond conventional in vivo and in vitro systems. The versatility of these methods has been particularly evident in areas such as environmental monitoring, advanced diagnostics, and biomolecular research. For environmental monitoring, QDs are employed as highly sensitive fluorescence-based sensors to detect a wide range of toxins, heavy metals, and pollutants [71]. Functionalized QDs conjugated with specific ligands have demonstrated exceptional efficacy in identifying contaminants such as mercury and arsenic in water with unprecedented sensitivity [72].

In the domain of biosensing, QDs serve as critical components in real-time and point-of-care diagnostic platforms. The high fluorescence stability and tuneable emission spectra enable the precise detection of biomolecules, such as glucose in blood samples and DNA sequences, during hybridization assays [73]. Furthermore, QDs have shown significant utility in food safety applications, where they are used to identify pathogens, toxins, and contaminants, providing a robust tool for ensuring food quality and safety [74]. Another notable application lies in Förster resonance energy transfer (FRET)-based studies, where QDs are integral in elucidating complex biological interactions. These include protein–protein interactions, enzyme activities, and nucleic acid dynamics [75]. The ability of QDs to act as both donor and acceptor molecules in FRET assays provides high spatial and temporal resolution, advancing our understanding of molecular biology [76].

In synthetic biology, QDs enable visualization and tracking of engineered biological systems, such as synthetic gene circuits. For example, studies have utilized QDs to monitor genetically engineered microorganisms in environmental contexts, demonstrating their capacity to integrate seamlessly into biological frameworks for dynamic tracking and analysis (Fig. 4) [77].

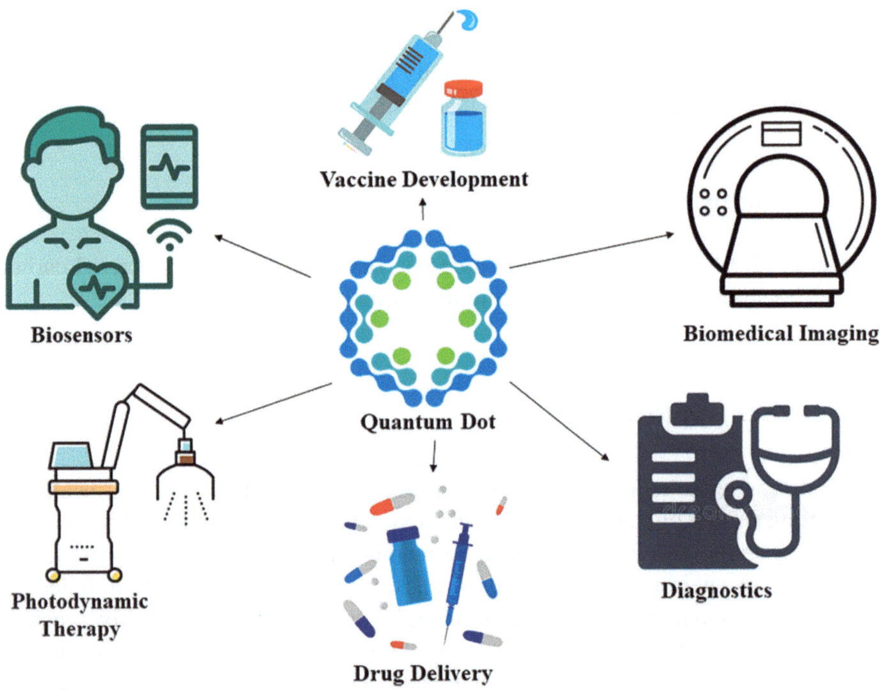

Fig. 4 Schematic representation of various applications of QDs in the biomedical field

6 Conclusion and Future Perspectives

QDs (QDs) have emerged as transformative tools in biology because of their exceptional optical and electronic properties, offering immense potential across a spectrum of applications. From in vitro diagnostics and imaging to in vivo cancer therapy and drug delivery, QDs have demonstrated unparalleled capabilities in enhancing precision, sensitivity, and efficiency. In addition to their role in biological systems, their role in environmental monitoring, synthetic biology, and biosensing further highlights their versatility.

The development of innovative QD variants, such as alloyed and core–shell structures, alongside advancements in surface modifications, has significantly expanded their biocompatibility and reduced toxicity, paving the way for safer and more effective applications. Photodynamic therapy, targeted drug delivery, and multimodal imaging are prime examples of how QDs are revolutionizing medical and therapeutic paradigms. Furthermore, their integration into fields such as proteomics, food safety, and environmental science underpins their multidisciplinary relevance.

Despite these advancements, challenges such as toxicity, stability, and scalability persist. Addressing these issues through eco-friendly synthesis methods, cadmium-free materials, and advanced functionalization strategies will be critical in fully harnessing the potential of QDs. Future directions include leveraging artificial intelligence for QD-based diagnostics and expanding its utility in personalized medicine and environmental sustainability.

Acknowledgement The authors would like to acknowledge the Wellcome Trust DBT India Alliance Intermediate Fellowship (IA/I/22/1/506240), SPARC-Ministry of Human Resource Development, Government of India (P2297), SERB-MATRICS, Government of India. (MTR/2021/000182). The authors also thank the Manipal School of Life Sciences, Manipal Academy of Higher Education, Manipal, India.

References

1. Ornes, S. (2016). The brighter the better: Quantum dots take a shine to biomedical applications. *Proceedings of the National Academy of Sciences, 113*(18), 4840–4842. https://doi.org/10.1073/pnas.1604839113
2. Pandey, A., Prasad, A., & Rai, V. (2020). Quantum dots: Promising nanosensors for diagnosis and monitoring of infectious diseases. *Analytica Chimica Acta, 1108*, 1–20. https://doi.org/10.1016/j.aca.2020.02.004
3. Ankireddy, S. R., Kim, D., & Kim, T. (2017). Quantum dot-based sensors for the detection of heavy metal ions: A review. *Sensors, 17*(8), 1881. https://doi.org/10.3390/s17081881
4. Sathe, T. R., Zou, J., & Leburton, J. P. (2022). Optical properties of quantum dots: The role of size and shape in emission characteristics. *Journal of Nanophotonics, 16*(2), 026005. https://doi.org/10.1117/1.JNP.16.026005
5. Bawendi, M. G., Steigerwald, M. L., & Brus, L. E. (1990). The quantum mechanics of larger semiconductor clusters ("quantum dots"). *Annual Review of Physical Chemistry, 41*(1), 477–496. https://doi.org/10.1146/annurev.pc.41.100190.002401

6. Ganguly, S., Das, P., Banerjee, S., & Das, N. C. (2019). Advancement in science and technology of carbon dot-polymer hybrid composites: A review. *Functional Composites and Structures, 1*(2), 022001. https://doi.org/10.1088/2631-6331/ab0c80
7. Pattantyus-Abraham, A. G., Kramer, I. J., Barkhouse, A. R., Wang, X., Konstantatos, G., Debnath, R., Levina, L., Raabe, I., Nazeeruddin, M. K., Grätzel, M., & Sargent, E. H. (2010). Depleted-heterojunction colloidal quantum dot solar cells. *ACS Nano, 4*(6), 3374–3380. https://doi.org/10.1021/nn100335g
8. Pisanic, T. R., II, Zhang, Y., & Wang, T. H. (2014). Quantum dots in diagnostics and detection: Principles and paradigms. *The Analyst, 139*(12), 2968–2981. https://doi.org/10.1039/c4an00294f
9. Soheyli, E., Behrouzi, S., Sharifirad, Z., & Sahraei, R. (2019). Multicolored type-I Ag-doped ZnCdS/ZnS core/shell quantum dots with intense emission. *Ceramics International, 45*(9), 11501–11507. https://doi.org/10.1016/j.ceramint.2019.03.019
10. Dai, X., Zhang, Z., Jin, Y., Niu, Y., Cao, H., Liang, X., Chen, L., Wang, J., & Peng, X. (2014). Solution-processed, high-performance light-emitting diodes based on quantum dots. *Nature, 515*(7525), 96–99. https://doi.org/10.1038/nature13829
11. Liu, Z., Lin, C., Hyun, B., Sher, C., Lv, Z., Luo, B., Jiang, F., Wu, T., Ho, C., Kuo, H., & He, J. (2020). Microlight-emitting diodes with quantum dots in display technology. *Light Science & Applications, 9*(1), 83. https://doi.org/10.1038/s41377-020-0268-1
12. Mahler, B., Spinicelli, P., Buil, S., Quelin, X., Hermier, J., & Dubertret, B. (2008). Towards nonblinking colloidal quantum dots. *Nature Materials, 7*(8), 659–664. https://doi.org/10.1038/nmat2222
13. Cheng, Y., Da Ling, S., Geng, Y., Wang, Y., & Xu, J. (2021). Microfluidic synthesis of quantum dots and their applications in biosensing and bioimaging. *Nanoscale Advances, 3*(8), 2180–2195. https://doi.org/10.1039/d0na00933d
14. Zorab, M. M., Mohammadjani, N., Ashengroph, M., & Alavi, M. (2022). Biosynthesis of quantum dots and their therapeutic applications in the diagnosis and treatment of cancer and SARS-CoV-2. *Advanced Pharmaceutical Bulletin, 13*(3), 411–422. https://doi.org/10.34172/apb.2023.065
15. Mao, C., Flynn, C. E., Hayhurst, A., Sweeney, R., Qi, J., Georgiou, G., Iverson, B., & Belcher, A. M. (2003). Viral assembly of oriented quantum dot nanowires. *Proceedings of the National Academy of Sciences, 100*(12), 6946–6951. https://doi.org/10.1073/pnas.0832310100
16. Hewa-Rahinduwage, C. C., Silva, K. L., Brock, S. L., & Luo, L. (2021). Quantum dot assembly driven by electrochemically generated metal-ion crosslinkers. *Chemistry of Materials, 33*(12), 4522–4528. https://doi.org/10.1021/acs.chemmater.1c00832
17. Bajwa, N., Mehra, N. K., Jain, K., & Jain, N. K. (2015). Pharmaceutical and biomedical applications of quantum dots. *Artificial Cells Nanomedicine and Biotechnology*, 1–11. https://doi.org/10.3109/21691401.2015.1052468
18. Sealy, C. (2019). Shape memory polymer responds to biological activity. *Materials Today, 25*, 6. https://doi.org/10.1016/j.mattod.2019.03.004
19. Le, N., & Kim, K. (2023). Current advances in the biomedical applications of quantum dots: Promises and challenges. *International Journal of Molecular Sciences, 24*(16), 12682. https://doi.org/10.3390/ijms241612682
20. Chinnathambi, S., Chen, S., Ganesan, S., & Hanagata, N. (2013). Silicon quantum dots for biological applications. *Advanced Healthcare Materials, 3*(1), 10–29. https://doi.org/10.1002/adhm.201300157
21. Shen, J., & Zhu, Q. (2022). Stability strategies of perovskite quantum dots and their extended applications in extreme environment: A review. *Materials Research Bulletin, 156*, 111987. https://doi.org/10.1016/j.materresbull.2022.111987
22. Clarke, M. T., Viscomi, F. N., Chamberlain, T. W., Hondow, N., Adawi, A. M., Sturge, J., Erwin, S. C., Bouillard, J. G., Tamang, S., & Stasiuk, G. J. (2019). Synthesis of super bright indium phosphide colloidal quantum dots through thermal diffusion. *Communications Chemistry, 2*(1). https://doi.org/10.1038/s42004-019-0138-z

23. Zhou, J., Yang, Y., & Zhang, C. (2015). Toward biocompatible semiconductor quantum dots: From biosynthesis and bioconjugation to biomedical application. *Chemical Reviews, 115*(21), 11669–11717. https://doi.org/10.1021/acs.chemrev.5b00049
24. Bera, D., Qian, L., Tseng, T.-K., & Holloway, P. H. (2010). Quantum dots and their multimodal applications: A review. *Materials, 3*(4), Article 4. https://doi.org/10.3390/ma3042260
25. Myers, K. A., & Janetopoulos, C. (2016). Recent advances in imaging subcellular processes. *F1000Research, 5*, F1000 Faculty Rev-1553. https://doi.org/10.12688/f1000research.8399.1
26. Yao, S., Hu, Y., & Li, G. (2014). A one-step sonoelectrochemical preparation method of pure blue fluorescent carbon nanoparticles under a high intensity electric field. *Carbon, 66*, 77–83. https://doi.org/10.1016/j.carbon.2013.08.044
27. Sun, Y.-P., Zhou, B., Lin, Y., Wang, W., Fernando, K. A. S., Pathak, P., Meziani, M. J., Harruff, B. A., Wang, X., Wang, H., Luo, P. G., Yang, H., Kose, M. E., Chen, B., Veca, L. M., & Xie, S.-Y. (2006). Quantum-sized carbon dots for bright and colorful photoluminescence. *Journal of the American Chemical Society, 128*(24), 7756–7757. https://doi.org/10.1021/ja062677d
28. Valizadeh, A., Mikaeili, H., Samiei, M., Farkhani, S. M., Zarghami, N., Kouhi, M., Akbarzadeh, A., & Davaran, S. (2012). Quantum dots: Synthesis, bioapplications, and toxicity. *Nanoscale Research Letters, 7*(1), 480. https://doi.org/10.1186/1556-276X-7-480
29. Karanikolos, G. N., Alexandridis, P., Itskos, G., Petrou, A., & Mountziaris, T. J. (2004). Synthesis and size control of luminescent ZnSe nanocrystals by a microemulsion–gas contacting technique. *Langmuir, 20*(3), 550–553. https://doi.org/10.1021/la035397+
30. Yaghini, E., Seifalian, A. M., & MacRobert, A. J. (2009). Quantum dots and their potential biomedical applications in photosensitization for photodynamic therapy. *Nanomedicine, 4*(3), 353–363. https://doi.org/10.2217/nnm.09.9
31. Gomase, A., Sangale, S., Mundhe, A., Gadakh, P., & Nikam, V. (2019). Quantum dots: Method of preparation and biological application. *Journal of Drug Delivery and Therapeutics, 9*(4-s), Article 4-s. https://doi.org/10.22270/jddt.v9i4-s.3333
32. Cao, Q., Feng, J., Chang, K. T., Liang, W., & Lu, H. (2024). Emerging opportunities of colloidal quantum dots for photocatalytic organic transformations. *Advanced Materials, 37*, e2409096. https://doi.org/10.1002/adma.202409096
33. Wegner, K. D., & Resch-Genger, U. (2024). The 2023 Nobel prize in chemistry: Quantum dots. *Analytical and Bioanalytical Chemistry, 416*(14), 3283–3293. https://doi.org/10.1007/s00216-024-05225-9
34. Ji, X., Wang, W., & Mattoussi, H. (2015). Controlling the spectroscopic properties of quantum dots via energy transfer and charge transfer interactions: Concepts and applications. *Nano Today, 11*(1), 98–121. https://doi.org/10.1016/j.nantod.2015.09.004
35. Alvelid, J. (2016). *Investigation of the photophysical properties of quantum dots for super-resolution imaging*. http://www.diva-portal.org/smash/record.jsf?pid=diva2:946275
36. La Rosa, M., Payne, E. H., & Credi, A. (2020). Semiconductor quantum dots as components of photoactive supramolecular architectures. *Chemistry Open, 9*(2), 200–213. https://doi.org/10.1002/open.201900336
37. Zhang, L., Xu, H., Zhang, X., Chen, X., Lv, Y., Zhang, R., Wang, L., Wu, R., Shen, H., & Li, L. S. (2024). Highly sensitive, stable InP quantum dot fluorescent probes for quantitative immunoassay through nanostructure tailoring and biotin–streptavidin coupling. *Inorganic Chemistry, 63*(10), 4604–4613. https://doi.org/10.1021/acs.inorgchem.3c04153
38. Kajani, A. A., Pouresmaeili, A., Mehrgardi, M. A., & Javanmard, S. H. (2024). Heteroatom-doped magneto-fluorescent carbon dots, a potent agent for multimodal imaging. *Scientific Reports, 14*(1), 29111. https://doi.org/10.1038/s41598-024-80531-x
39. Rees, K., Darwish, G. H., & Algar, W. R. (2023). Dextran-functionalized supernanoparticle assemblies of quantum dots for enhanced cellular immunolabelling and imaging. *ACS Applied Materials & Interfaces, 15*(15), 18672–18684. https://doi.org/10.1021/acsami.3c00861
40. Praseetha, P. K., Litany, R. I. J., Alharbi, H. M., Khojah, A. A., Akash, S., Bourhia, M., Mengistie, A. A., & Shazly, G. A. (2024). Green synthesis of highly fluorescent carbon quan-

tum dots from almond resin for advanced theranostics in biomedical applications. *Scientific Reports, 14*(1), 24435. https://doi.org/10.1038/s41598-024-75333-0
41. Upadhyaya, L., Verma, S., Verma, A., & Mishra, P. R. (2024). Nitrogen-doped carbon quantum dots as an efficient nanocarrier for curcumin delivery. *Journal of Drug Delivery Science and Technology, 85*, 105598. https://doi.org/10.1016/j.jddst.2023.105598
42. Upadhyaya, A. K., Agarwala, P., Sharma, C., & Sasmal, D. K. (2025). Synthesis and characterization of N-doped carbon quantum dots and its application for efficient delivery of curcumin in live cell. *ChemPhysChem, 26*(6), e202400855. https://doi.org/10.1002/cphc.202400855
43. Khobrekar, P. P., Zalmi, G. A., Raiturker, A. P., Jadhav, R. W., Ganguly, A., D'Costa, A., Bugde, S. T., & Bhosale, S. V. (2025). Citric acid functionalized neomycin carbon dots for cytotoxicity and sensing application. *Journal of Molecular Structure, 1323*, 140769. https://doi.org/10.1016/j.molstruc.2024.140769
44. Kadyan, P., Kumar, M., Tufail, A., Ragusa, A., Kumar Kataria, S., & Dubey, A. (2025). Microwave-assisted green synthesis of fluorescent graphene quantum dots (GQDs) using Azadirachta indica leaves: Enhanced synergistic action of antioxidant and antimicrobial effects and unveiling computational insights. *Materials Advances, 6*(2), 805–826. https://doi.org/10.1039/D4MA00843J
45. Khan, B., Zhang, J., Durrani, S., Wang, H., Nawaz, A., Durrani, F., Ye, Y., Wu, F.-G., & Lin, F. (2024). Carbon-dots-mediated improvement of antimicrobial activity of natural products. *ACS Applied Materials & Interfaces, 16*(36), 47257–47269. https://doi.org/10.1021/acsami.4c09689
46. Kumar, M., Chinnathambi, S., Bakhori, N., Abu, N., Etezadi, F., Thangavel, V., Packwood, D., Sivaniah, E., & Pandian, G. N. (2024). Biomass-derived carbon dots as fluorescent quantum probes to visualize and modulate inflammation. *Scientific Reports, 14*(1), 12665. https://doi.org/10.1038/s41598-024-62901-7
47. Cho, S., Kim, H., Song, D., Jung, J., Park, S., Jo, H., Seo, S., Han, C., Park, S., Kwon, W., & Han, H. (2024). Insights into glucose-derived carbon dot synthesis via Maillard reaction: From reaction mechanism to biomedical applications. *Scientific Reports, 14*(1), 31325. https://doi.org/10.1038/s41598-024-82767-z
48. Ramajayam, R., Selvam, K., Babu, S. G., & Chen, S.-M. (2024). Boron and nitrogen co-doped carbon dots for selective fluorimetric detection of Au^{3+} ions. *Journal of Photochemistry and Photobiology A: Chemistry, 437*, 114542. https://doi.org/10.1016/j.jphotochem.2022.114542
49. Ramajayam, K., & Palaniappan, A. (2025). Facile and green synthesis of highly fluorescent boron and nitrogen-doped carbon dots (BN-CDs) for the fluorimetric detection of Au3+ ions in aqueous media and in in vitro cell imaging. *Colloids and Surfaces A: Physicochemical and Engineering Aspects, 709*, 136062. https://doi.org/10.1016/j.colsurfa.2024.136062
50. Li, Y., Zhang, X., Wang, L., & Liu, J. (2024). Quantum dot-based biosensors for early cancer detection: Recent advances and future perspectives. *Biosensors and Bioelectronics, 233*, 115480. https://doi.org/10.1016/j.bios.2023.115480
51. Li, Y., Dang, G., Rizwan Younis, M., Cao, Y., Wang, K., Sun, X., Zhang, W., Zou, X., Shen, H., An, R., Dong, L., & Dong, J. (2023). Peptide functionalized actively targeted MoS2 nanospheres for fluorescence imaging-guided controllable pH-responsive drug delivery and collaborative chemo/photodynamic therapy. *Journal of Colloid and Interface Science, 639*, 302–313. https://doi.org/10.1016/j.jcis.2023.02.027
52. Ahmadi Kamalabadi, M., Ostadebrahimi, H., Koosha, F., Fatemidokht, A., Menbari Oskuie, I., Amin, F., & Shiralizadeh Dezfuli, A. (2024). Gd-GQDs as nanotheranostic platform for the treatment of HPV-positive oropharyngeal cancer. *Medical Oncology, 41*(8), 205. https://doi.org/10.1007/s12032-024-02431-4
53. Pons, T., Pic, E., Lequeux, N., Cassette, E., Bezdetnaya, L., Guillemin, F., Marchal, F., & Dubertret, B. (2010). Cadmium-free CuInS2/ZnS quantum dots for sentinel lymph node imaging with reduced toxicity. *ACS Nano, 4*(5), 2531–2538. https://doi.org/10.1021/nn901421v

54. Liu, Q., Guo, B., Rao, Z., Zhang, B., & Gong, J. R. (2013). Strong two-photon-induced fluorescence from photostable, biocompatible nitrogen-doped graphene quantum dots for cellular and deep-tissue imaging. *Nano Letters, 13*(6), 2436–2441. https://doi.org/10.1021/nl400368v
55. Jena, B. R., Pattnaik, G., & Samal, H. B. (2024). Enzyme-responsive micro/nano drug delivery. In *Smart micro- and nanomaterials for drug delivery*. CRC Press.
56. Zhang, H., Liu, R., Wang, Y., Chen, W., & Li, J. (2023). Red-emitting carbon dots with deep tissue penetration for highcontrast bioimaging applications. *Carbon, 211*, 118024. https://doi.org/10.1016/j.carbon.2023.118024
57. Fan, J., Zhang, J., Huang, W., Zhou, W., Zhou, L., Pang, Q., Chen, P., Wu, Z., Mo, F., Xiao, Q., & Zhang, X. (2023). Efficient red-emitting carbon dots for bioimaging and spectral engineering by polyetherimide modification. *Dyes and Pigments, 218*, 111492. https://doi.org/10.1016/j.dyepig.2023.111492
58. Shu, G., Jiang, T., Zhang, X., & Zhao, H. (2024). Gadolinium and Ytterbium-doped carbon dots for fluorescence/magnetic resonance/computed tomography multimodal imaging. *Chemical Engineering Journal, 494*, 153086. https://doi.org/10.1016/j.cej.2024.153086
59. Matea, C. T., Mocan, T., Tabaran, F., Pop, T., Mosteanu, O., Puia, C., Iancu, C., & Mocan, L. (2017). Quantum dots in imaging, drug delivery and sensor applications. *International Journal of Nanomedicine, 12*, 5421–5431. https://doi.org/10.2147/IJN.S138624
60. Dar, M. S., & Sahu, N. K. (2024). Graphene quantum dot-crafted nanocomposites: Shaping the future landscape of biomedical advances. *Discover Nano, 19*(1), 79. https://doi.org/10.1186/s11671-024-04028-2
61. Song, Y., Chen, Y., Wang, Z., Zhang, L., & Liu, X. (2024). Enzymatic cascade nanoreactors based on carbon dots for tumor-targeted self-enhancing drug delivery and synergistic therapy. *ACS Nano, 18*(3), 3456–3469. https://doi.org/10.1021/acsnano.3c12345
62. Song, L., Zhao, Q., Feng, S., He, Y., Li, Y., Wang, S., & Zhang, J. (2025). Depolymerizable enzymatic cascade nanoreactor for self-enhancing targeting synergistic tumor therapy. *Advanced Functional Materials, 35*(4), 2414121. https://doi.org/10.1002/adfm.202414121
63. Vayalakkara, N., Thomas, S., Pillai, A., & Menon, D. (2024). Immunomodulatory agent-loaded reduced graphene oxide quantum dots for enhanced photothermal therapy of glioblastoma. *Biomaterials Advances, 161*, 213456. https://doi.org/10.1016/j.bioadv.2024.213456
64. Araújo-Silva, F. D., Lima, T. S., Carvalho, M. C., & Andrade, G. R. S. (2024). Nanographene oxide conjugates for enhanced photodynamic therapy: High drug-loading efficiency and tumor-specific targeting. *Journal of Photochemistry and Photobiology B: Biology, 251*, 112423. https://doi.org/10.1016/j.jphotobiol.2024.112423
65. Araújo-Silva, H., Martins, B., Cerqueira, F., Gonçalves, H., Gomes, A. C., & Lúcio, M. (2025). Novel nanographene oxide conjugates as stimuli-responsive theranostic tool for cancer. *Materials Today Chemistry, 44*, 102552.
66. Wang, Y., Du, M., Wang, J., Bai, Z., Cui, C., Tong, J., Liu, Y., Guo, S., Zhang, W., Wu, X., & Li, B. (2025). Carbon dots-cisplatin nano drug delivery system induces the death of oral tongue squamous cell under self-targeting chemical/photodynamic combined therapy. *Colloids and Surfaces A: Physicochemical and Engineering Aspects, 704*, 135511. https://doi.org/10.1016/j.colsurfa.2024.135511
67. Yan, X., An, N., Zhang, Z., Qiu, Q., Yang, D., Wei, P., Zhang, X., Qiu, L., & Guo, J. (2024). Graphene oxide quantum dots-preactivated dental pulp stem cells/GelMA facilitates Mitophagy-regulated bone regeneration. *International Journal of Nanomedicine, 19*, 10107–10128. https://doi.org/10.2147/IJN.S480979
68. Kanemaru, K., Panniello, M., Calì, B., & Nardin, C. (2024). Editorial: Lighting up cellular dynamics with fluorescent biosensors: Design and applications in pathophysiology. *Frontiers in Physiology, 15*, 1517919. https://doi.org/10.3389/fphys.2024.1517919
69. Filali, S., Pirot, F., & Miossec, P. (2020). Biological applications and toxicity minimization of semiconductor quantum dots. *Trends in Biotechnology, 38*(2), 163–177. https://doi.org/10.1016/j.tibtech.2019.07.013

70. Jiang, P.-L., Hong, Y.-Y., Yang, L., Lin, H.-J., Huang, C.-C., Chen, Y.-H., Lin, C.-H., & Chen, Y.-C. (2024). Comprehensive evaluation of the nephrotoxicity of carbon quantum dots: Effects of the surface charge. *Chemosphere, 367*, 143604. https://doi.org/10.1016/j.chemosphere.2024.143604
71. Freire, M. S., Silva, H. J. B., Albuquerque, G. M., Monte, J. P., Lima, M. T. A., Silva, J. J., Pereira, G. A. L., & Pereira, G. (2024). Advances on chalcogenide quantum dots-based sensors for environmental pollutants monitoring. *Science of the Total Environment, 931*, 172848. https://doi.org/10.1016/j.scitotenv.2024.172848
72. Sun, M., Yang, L., Jose, P., Wang, L., & Zweit, J. (2013). Functionalization of quantum dots with multidentate zwitterionic ligands: Impact on cellular interactions and cytotoxicity. *Journal of Materials Chemistry B, 1*(44), 6137–6146. https://doi.org/10.1039/C3TB20894J
73. Vannoy, C. H., Tavares, A. J., Noor, M. O., Uddayasankar, U., & Krull, U. J. (2011). Biosensing with quantum dots: A microfluidic approach. *Sensors (Basel, Switzerland), 11*(10), 9732–9763. https://doi.org/10.3390/s111009732
74. Mattoussi, H., Palui, G., & Na, H. B. (2012). Luminescent quantum dots as platforms for probing in vitro and in vivo biological processes. *Advanced Drug Delivery Reviews, 64*(2), 138–166. https://doi.org/10.1016/j.addr.2011.09.011
75. Day, R. N. (2014). Measuring protein interactions using Förster resonance energy transfer and fluorescence lifetime imaging microscopy. *Methods, 66*(2), 200–207. https://doi.org/10.1016/j.ymeth.2013.06.017
76. Song, Y., Gao, D., Kim, H. S., Qu, C., Kang, J., Zhu, Y., Liu, Z., Guo, J., Xu, L., & Lee, C. S. (2014). Fabrication of silicon carbide quantum dots via chemical-etching approach and fluorescent imaging for living cells. *Materials Sciences and Applications, 5*(4), Article 4. https://doi.org/10.4236/msa.2014.54022
77. Rafeeq, H., Afsheen, N., Rafique, S., Arshad, A., Intisar, M., Hussain, A., Bilal, M., & Iqbal, H. M. N. (2023). Genetically engineered microorganisms for environmental remediation. *Chemosphere, 310*, 136751. https://doi.org/10.1016/j.chemosphere.2022.136751

Sensing Applications of Perovskite Quantum Dots

Priyankamoni Saikia, Asfi Ahmed, Sanjeev Pran Mahanta, and Swapan K. Dolui

1 Introduction

A sensor aims to detect changes in the physical environment upon interaction with an analyte and converts these changes into readable and reliable signals [1]. Volatile compounds such as carbon monoxide, ammonia, alcohol, ether, ester, metal ions, explosives, and different pesticides not only contribute to air pollution in the environment but also have adverse effects on human health, such as respiratory diseases [2], central nervous system disorders [3], skin allergies [4], and fatigue [5]. Rapid, selective, sensitive detection is highly important in many pharmaceutical and industrial processes and food industries as well as in medical and military applications [6]. Long-term exposure to alcohol vapor and pesticides can stimulate severe health disturbances, such as nerve disease, lung irritation, disturbance in eyesight, and respiratory disorders, and even lead to death in many cases [7, 8]. There is a demanding need to detect and identify the nature of different target analytes. To date, various spectroscopic techniques, including Chromatography, Nuclear Magnetic Resonance, Ultraviolet–Visible Spectroscopy, Fourier Transform Infrared Spectroscopy (FTIR), Mass Spectrometry and Raman Spectroscopy, have been used to detect different types of analytes [9–15]. However, there are several distinct limitations to the use of those spectroscopic techniques, such as slow responses, low selectivity, high cost, and sensitive complicated operations [16]. To resolve these

P. Saikia · A. Ahmed · S. K. Dolui (✉)
Department of Chemical Sciences, Tezpur University, Napaam, Tezpur, Assam, India
e-mail: dolui@tezu.ernet.in

S. P. Mahanta
Department of Chemical Sciences, Tezpur University, Napaam, Tezpur, Assam, India

Centre for Multidisciplinary Research, Tezpur University, Napaam, Tezpur, Assam, India

S. P. Kabekkodu et al. (eds.), *Functional Quantum Dots*, Lecture Notes in Nanoscale Science and Technology 40,
https://doi.org/10.1007/978-3-031-98219-4_9

problems, the development of a new generation of sensors demands high selectivity, rapid performance, and a simple portable target detection system.

Several semiconductor materials, including TiO_2 [17] and ZnO [18], have been used as sensing detectors because of their large surface area, biocompatibility, etc. However, the notable drawback of those materials is that a high-temperature process is involved in their manufacture. To overcome those challenges in the preparation of sensory materials, finding new semiconductor materials for sensing is currently a relevant research topic where enormous investments are essential.

Quantum dots have emerged as new materials with various advantageous properties. These materials have dimensions of 5–10 nm and feature excellent physical properties, high photoluminescence quantum yields (PLQYs), size-tunable emission wavelengths, etc. [19]. Owing to their different superior characteristics, quantum dots have wide applications in different fields, such as LEDs, sensors, photodetectors, and solar cells [20].

In the field of material chemistry, perovskite quantum dots are emerging semiconductor materials with various photophysical and electronic advantages [19]. Tsutomu Myaska and coworkers first reported inorganic and organic perovskites as light absorbers in dye-sensitized solar cells [21]. The perovskite quantum dot materials are promising candidates for solar cells, field-effect transistors, light-emitting diodes, gas sensors, etc. [20] (Fig. 1).

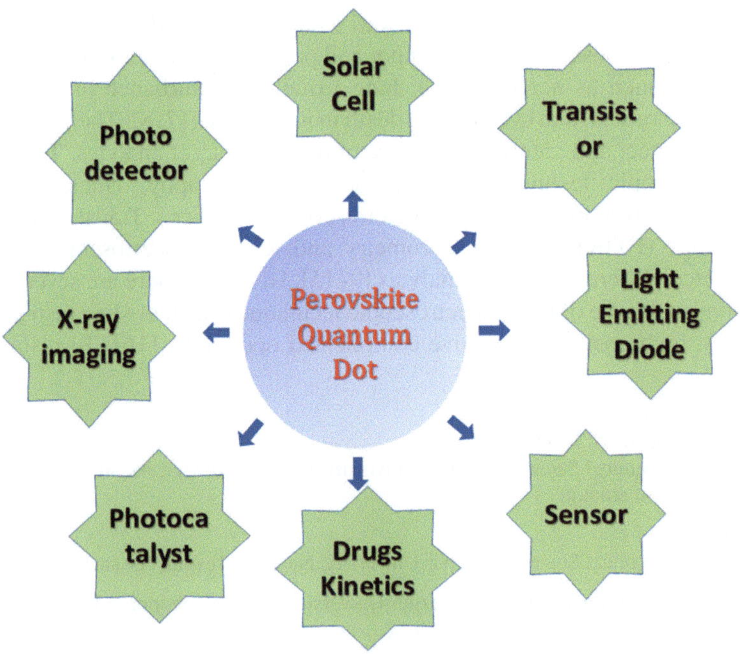

Fig. 1 Different applications of Perovskite Quantum Dots for energy conversion and storage

Perovskites emerge as materials with the molecular formula ABX_3 (where A & B = cations and X = halide). In the cubic structure ABX_3, the six halogen atoms (X) combine to form an octahedral structure, and metal B is located at the center. However, cation A occupied the center of the eight octahedra (Fig. 2). On the basis of their composition, perovskites can be classified into three categories: organic metal halide perovskites, inorganic oxide perovskites, and alkaline metal halide perovskites. The perovskite quantum dot material features an absorption coefficient that is 10 times greater than those of other light-absorbing materials. Additionally, the high quantum yield and increased luminescence intensity are major factors in the application of perovskite crystals in applications of optoelectronic devices [22]. The material is sensitive to many polar molecules because of its unique crystal structure [23].

In the presence of various analytes, some structural changes in the perovskite crystal lattice occur, which makes them efficient sensors for detecting various analytes. The interaction between the sensor and analyte may occur through H-bonding interactions, Fourier transform electron transfer, intermolecular charge transfer, etc. In this chapter, we summarize various methods or techniques that evolved during the sensing process for different analytes by using perovskite quantum dot materials (Table 1). The detection approaches have been classified into different categories on the basis of the physical and chemical processes involved. The selectivity towards different analyte molecules of that perovskite quantum dot material is improved by doping with other foreign metals and making some composites with some additives. Finally, a conclusion focused on sensing processes and future challenges while using perovskite quantum dot materials is also discussed in this chapter.

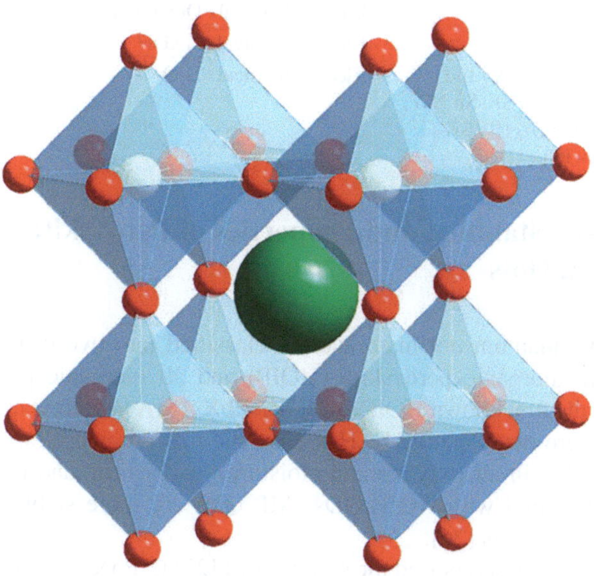

Fig. 2 Structure of a perovskite (copyright permitted)

Table 1 Comparison of different methods of sensing different analytes by perovskite quantum dot materials

Sl. no.	Perovskite quantum dot	Sensing method	Analyte	References
1	$CsPbBr_3$ PQDs/cellulose composite	Colorimetric	(i) Cl^- (ii) I^-	[45]
2	$CH_3NH_3PbBr_3$	Fluorometric	F^-	[58]
3	$CsPbBr_3/CsPbI_3$	Fluorometric	Rhein	[59]
4	$CsPbBr_3$	Fluorometric	Polar content in edible oil	[46]
5	$CsPbBr_3$ perovskite quantum dot decorated ZIF-8 MOF	Fluorometric	(i) 4-nitroaniline (ii) rhodamine blue	[60]
6	$CsPbBr_3$	Fluorometric	Rhodamine 6G	[7]
7	MIP@$MAPbBr_3$	Fluorometric	Benzo-(α)-pyrene (BaP)	[61]
8	Hydroxypropyl chitosan-$Cs_3Bi_2Cl_9$	Fluorometric	Chromium [Cr(VI)]	[62]
9	$CsPbI_3$	Fluorometric	Carcinoembryonic antigen and neuron specific enolase	[63]
10	$CsPbBr_3$/molecularly imprinted polymers (MIPs)	Fluorometric	Omethoate	[48]
11	$CsPbBr_3$	Fluorometric	Cu^{2+}	[49]
12	$Cs_3Bi_2Br_9/Eu^{3+}$	Fluorometric	Cu^{2+}	[50]
13	$CH_3NH_3PbBr_3$	Fluorometric	Hg^{2+}	[51]
14	$CsPbBr_3$/APTES	Fluorometric	Tetracycline	[52]
15	$MAPbBr_3$	Fluorometric	Organochlorine pesticides	[53]
16	$CsPbBr_{1.5}I_{1.5}/TiO_2$	Electrochemical	Dopamine	[57]
17	$CsPbBr_3/ZnO$	Electrochemical	NADH	[58]
18	$CsPbBr_3$ (PVSK NW) and $CsPb_{1-x}Sn_xBr_3$ (PVSK-Sn NW) and $CsPb_{1-x}Ge_xBr_3$ (PVSK-Ge NW) nanowires	Electrochemical	H_2O_2	[59]

2 Synthesis, Stability, and Properties of Perovskite Quantum Dots

The perovskite quantum dot materials are classified into two distinct categories: hybrid organic–inorganic perovskite (HOIP) and all inorganic perovskite (AIP) [20]. In 1987, the first known HOIP QDs, $MAPbX_3$ (MA ¼ methylammonium; X ¼ Cl, Br or I), were developed [24]. However, the main drawback of HOIP QDs is their poor stability in air (oxygen and moisture), under heat, and under ultraviolet light [25]. Compared with HOIP QDs, AIP QDs are more stable where cesium replaces organic cations [26].

Protesescu et al. first reported the synthesis of [27] $CsPbX_3$ (X ¼ Cl, Br or I) QDs through a hot-injection method with stable and widely tunable photoluminescence

(QY of 50–90%). Li et al. designed a ligand-assisted reprecipitation (LARP) method to produce high-QY perovskite QDs while using two precursor solutions to produce supersaturated precipitation at room temperature [28]. Considering the toxicity of lead, the less toxic metal tin was subsequently used in place of Pb in the $CsPbX_3$ QDs. The first reported $CsSnX_3$ perovskite QDs were produced by Jellicoe et al. [29] through a hot-injection method with a low quantum yield of 0.14%. Additionally, Zhang et al. [30] used trivalent antimony (Sb^{3+}) to replace Pb^{2+} to develop brightly luminescent $Cs_3Sb_2Br_9$ QDs, with a quantum yield of 46%.

To increase the stability of perovskite QDs, a surface coating was used. Zhou et al. [31] designed perovskite QDs in a PVDF membrane with high PL efficiency in water and UV light. MOF frameworks [32] have also been used to increase the stability of perovskite QDs. Additionally, different surfactants [33], ligands [34], and composite materials [35] are used as encapsulation techniques to improve the stability of perovskite QDs.

Perovskite QDs exhibit a quantum confinement effect [36, 37]. Compared with bulk or single crystals, perovskite QDs have advantageous optical and electronic properties because of their low trap densities, high surface-to-volume ratio, and versatile functionalization [38]. The PLQY properties of colloidal perovskite QDs are more than 90%, making them appropriate candidates for application as optical sensors. The crystalline structures of perovskite QDs can be synthetically tuned. The luminescence properties of perovskite QDs exhibit brighter and narrower photoluminescence (PL) than organic polymers, dyes, and phosphors. The perovskite QDs feature a tenable spectral range by tuning both halide compositions (Fig. 3).

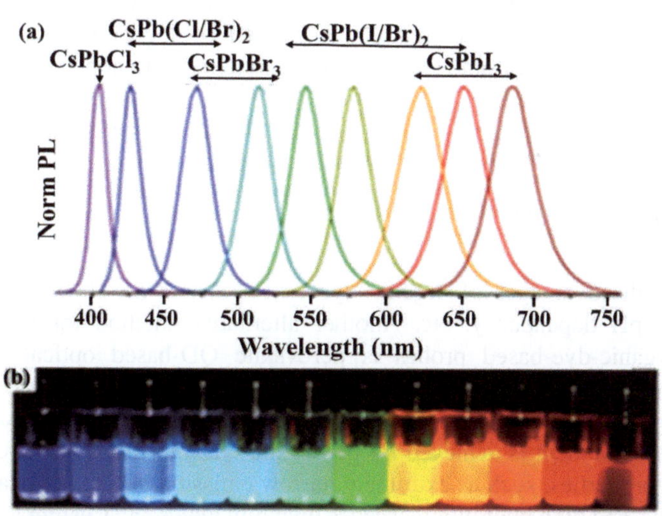

Fig. 3 (**a**) Emission spectra of a $CsPbX_3$ colloidal solution in toluene upon excitation at 365 nm (λ_{ex} = 400 nm for all, but 350 nm for $CsPbCl_3$); (**b**) digital photographs of the respective dispersions upon irradiation with a UV lamp (λ = 365 nm)

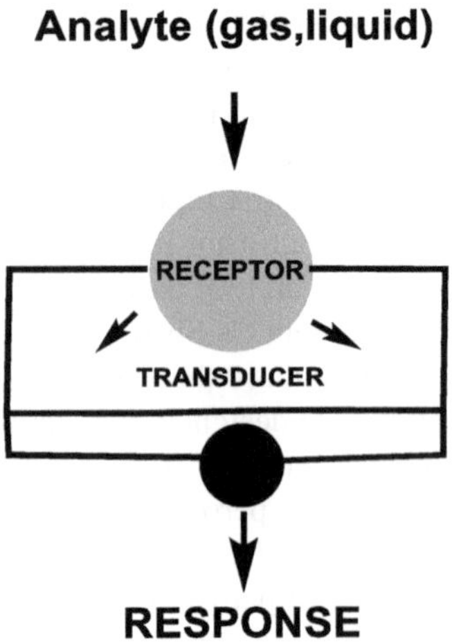

Fig. 4 Chemical sensor: Receptor and transducer

3 Chemical Sensors Based on Perovskite QDs

Conventionally, a chemical sensor is composed of two functional units: a receptor part and a transducer part (Fig. 4). The receptor part interacts with a particular analyte selectively and reversibly, and the transducer part converts the respective physical and chemical changes into some analytical signals (Fig. 4) [39, 40]. The nature of the receptor parts depends on the identity of the target analyte. The interactions between the receptor molecule/material and analytes are mostly noncovalent in nature, leading to physical changes. However, in certain systems, the interaction between the analyte and receptor triggers some chemical reactions leading to certain noticeable changes that are recognized by the transducer component. However, for long-term applications and reuse purposes, systems with noncovalent reversible interactions are preferred.

4 Optical Sensors

The major disadvantages of organic dye-based sensors are their self-quenching properties, pH dependency, etc. Another alternative method for sensor probes against organic-dye-based probes is perovskite QD-based optical sensors. In perovskite-QD-based sensors, the quenching of their luminescence properties occurs due to the binding of analytes on the surface of the perovskite. Additionally, quantum confinement is achieved by perovskite QDs, resulting in excellent optoelectronic properties, such as extremely narrow emission bands, long carrier diffusion lengths, and high photoluminescence quantum yields etc. Cesium lead bromide ($CsPbBr_3$) QDs have a higher quantum yield of up to 90% with narrow emission widths of 12 nm [27], which increases the number of parameters that can be utilized as optical sensors with a lower detection limit.

The interaction mechanisms have been rationalized as Forster resonance energy transfer (FRET), photoinduced electron transfer (PET), recombination, etc., in various studies (Fig. 5) [41, 42].

When a perovskite compound reacts with an analyte, its cation binds with the B-site of the fluorophores, creating some changes in its optical properties (Fig. 6). In the inorganic perovskite $CsPbBr_3$, the emission wavelength can be altered by exchanging the halide ion.

In the perovskite lattice, the arrangement of the valence band of the donor and the conduction band of the acceptor moieties can result in the same alignment. Owing to this phenomenon, perovskite materials are compatible with electron transfer between the sensing probe and the analyte. Consequently, some significant changes in the physical and chemical properties of the materials occur, and the resulting fluctuations help in the detection of analytes. During the FRET mechanism, spectral overlapping takes place between the PL emission spectra of the fluorophore and the UV–Vis absorbance spectra of the acceptor within a distance of 10–100 Å. In this mechanism, the energy of the excited donor is transferred to an acceptor molecule through a nonradiative process (Fig. 7). The process depends on the distance between the fluorophore and the analyte.

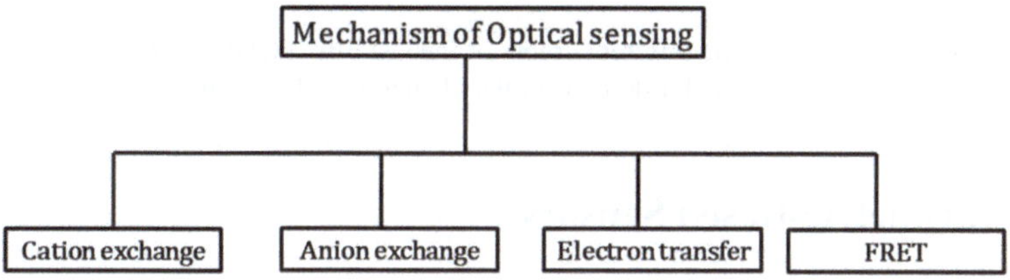

Fig. 5 Schematic diagram of the mechanism of perovskite QDs as optical sensors

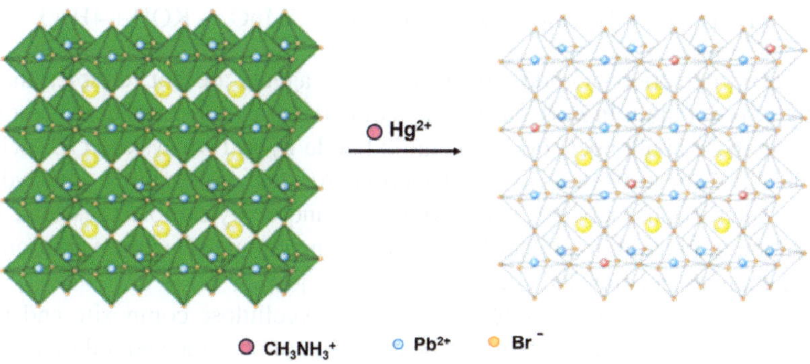

Fig. 6 Metal cation exchange mechanism

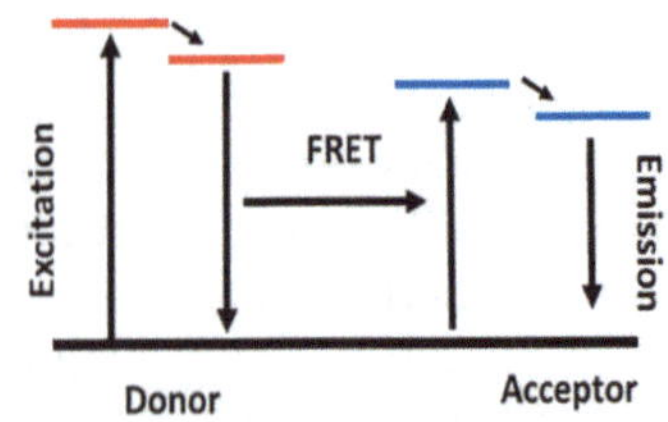

Fig. 7 FRET mechanism

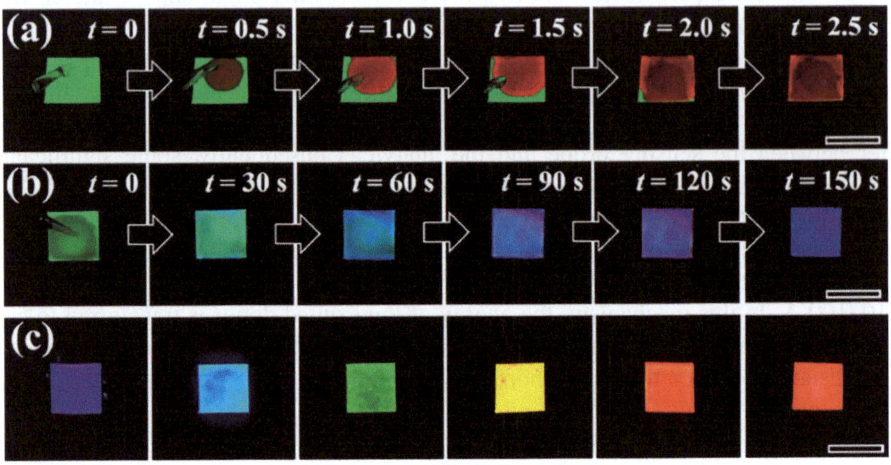

Fig. 8 Photographs showing the different colors of the $CsPbBr_3$ PQD/cellulose composites upon exposure to (**a**) iodide and (**b**) chloride and (**c**) the change in color from yellow to red under the gradual addition of I^- ions

5 Colorimetric-Based Sensors

Upon interaction with the analyte, a colorimetric sensor changes color, which may be attributed to structural changes. Most colorimetric methods use several oxidizing agents, such as potassium permanganate and cerium ammonium nitrate [43].

$$H_3CH_2OH + 4KMnO_4 \rightarrow 3CH_3COOH + 4MnO_2 + KOH + 4H_2O \quad (1)$$

However, detection methods using oxidizing agents are not reliable, and their sensitivity is not high. In most cases, they could demonstrate nonspecific results. To overcome these challenges, organometallic halide hybrid perovskite QDs with the formula ABX_3 have been used for detection. Anionic interactions and hydrogen bonding interactions between perovskite QDs and analyte molecules lead to the degradation of perovskite molecules. The resulting crystal structure variations lead to some changes in color and emission properties [44].

Park et al. synthesized a stable $CsPbBr_3$ PQD/cellulose composite and further applied it as a colorimetric sensor for the detection of chloride and iodide ions [45]. Through an anion exchange reaction, the ionic composition of the sensor rapidly changes. The perovskite film showed color changes from green to blue and red in the presence of the analyte chloride to iodide, and the quenching of the emission peak was due to changes in the perovskite crystal structure (Fig. 8). The sensor device can detect analytes with detection limits of 2.56 and 4.11 mM for iodine and chlorine ions, respectively. The size of the synthesized particles was approximately 15 nm. The synthesized $CsPbBr_3$ PQDs/cellulose composite was able to detect the analyte within 5 s in tap water immersed with 1.0 M I^- and Cl^- ions. The real-time analysis proves that the sensor is a useful platform in biomedical applications.

6 Fluorescence-Based Sensors

In recent years, fluorescence-based detector systems have gained widespread attention because of their low cost, easy operation methods, and low detection limits. Upon interaction with the target analytes, the luminescence peak intensity of a fluorescence sensor may be quenched or enhanced. On the basis of the change in luminescence properties, the fluorophore can detect the analyte. Hybrid perovskite QD materials have excellent photoluminescence (PL) properties and hence are suitable materials for sensor applications.

$CsPbBr_3$, a lead halide perovskite quantum dot, has been reported as a fluorescent probe for the detection of alcohols in edible oils [46]. Conventionally, lead halide perovskite molecules are not stable in polar environments. This instability makes lead halide perovskites appropriate for polarity sensors. The binding force between perovskites and the ligands (oleylamine or oleic acid) used in surface capping is not strong enough, and the ligand can easily be desorbed. In the presence of polar molecules, the protective layer of the perovskite molecule can be destroyed. This creates some defects in some nonradioactive recombination paths. Some quenching behaviors of the perovskite solution occur in the presence of polar environments. Typically, $CsPbBr_3$ is green when dispersed in toluene. However, with the addition of a polar solvent, the color changes due to the change in the crystal structure (from orthorhombic to tetragonal). With increasing polarity content (Fig. 9), the

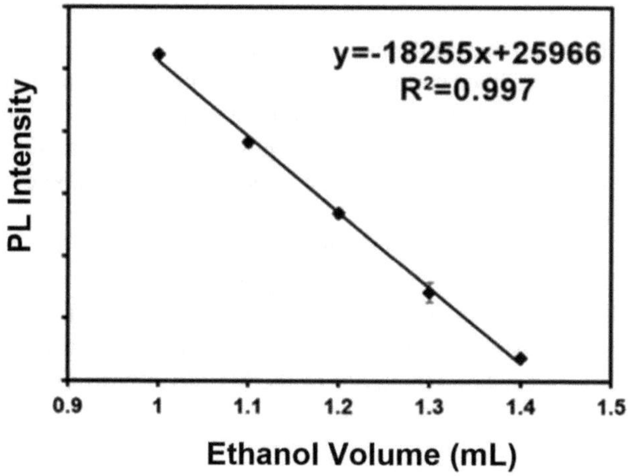

Fig. 9 The linear relationship between the intensity of PL absorbance and the volume of ethanol [47] (copyright permitted)

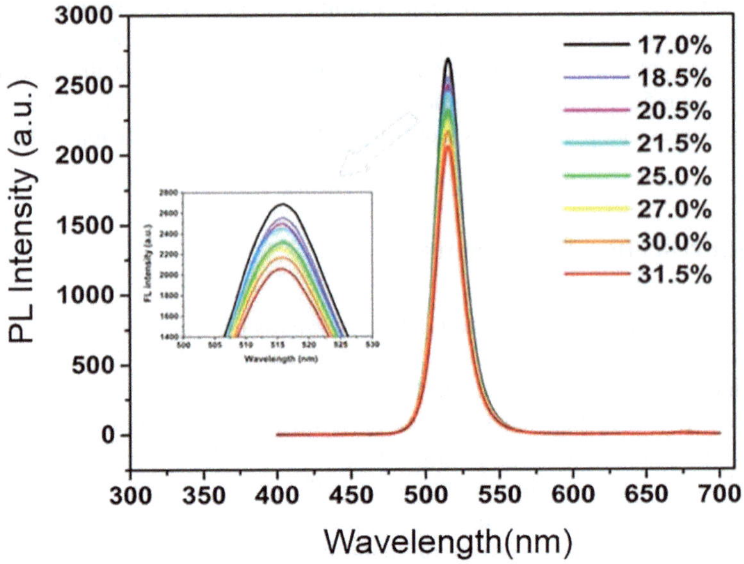

Fig. 10 Changes in the fluorescence spectra of $CsPbBr_3$ with the addition of edible oil (copyright permitted)

intensity of the luminescence peak decreases proportionally (Fig. 10). To calculate the effect of the polar solvent on the perovskite, 1 mL of a mixed solution of perovskite quantum dots with toluene was used. Various amounts of alcohol are added to the perovskite solution. However, for total polarity measurement (TPM) calculations in edible oil, a paper-based sensor device was fabricated. The correlation of the fluorescence method with the established method makes the sensor device feasible for TPM content in edible oils. Additionally, a quenching effect was observed for alcohols present in edible oils such as olive oil, sunflower oil, and soybean oil. The synthesized quantum dots exhibit a cubic structure with an average size of 10 nm.

Wang et al. reported the synthesis of $CsPbBr_3$ perovskite QDs encapsulated in a monolithic polystyrene fiber membrane through an electrospinning method for the detection of rhodamine 6G, with an ultralow detection limit of 0.01 ppm in aqueous and ethanol media [47]. The sensor device shows good recyclability and excellent potential for biomedical applications. The sensing mechanism takes place through the FRET electron transfer process.

Huang et al. reported highly stable luminescent perovskite ($CsPbBr_3$) quantum dots combined with molecularly imprinted polymers (MIPs) and further applied them for the detection of omethoate (OMT), with a detection limit of 18.8 ng/mL [48]. The designed sensor was successfully applied in the detection of OMT in vegetable and soil samples. The selectivity of the MIP@$CsPbBr_3$ QD composites was tested with several similar organophosphorus (OP) insecticides, such as dimethoate, dichlorvos, phoxim, chlorpyrifos, and acetamiprid (organochlorine pesticide). Among all OP pesticides, the sensor probe shows high selectivity towards OMT. The quenching process takes place through a dynamic quenching mechanism where in the excited state, the fluorescence substance collides with the quencher molecule and returns to the ground state. During the collision process, an energy-transfer mechanism or a charge-transfer mechanism takes place. The mechanism was further proven by time-resolved photoluminescence (TRPL) analysis, where after adding the analyte OMT, the average lifetime of the sensor probe decreased from 27.8 to 23.8 ns.

Furthermore, Seng et al. reported a colloidal solution of perovskite $CsPbBr_3$ perovskite QDs for the detection of Cu^{2+} in a nonpolar solvent [49]. In the reported perovskite crystal structure, the valence band maxima were dominated by Br^-, and the conduction band minima were contributed by Pb^{2+}. With the gradual addition of Cu^{2+} ions, a new state is created near the valence band maxima of the perovskite crystal structure. Upon excitation, the hole is occupied by this new state. As a result, a nonradiative recombination pathway is responsible for the quenching of the luminescence peak.

Additionally, Ding and coworkers synthesized an eco-friendly nonlead perovskite, $Cs_3Bi_2Br_9$. Later, they doped the synthesized perovskite with Eu^{3+} [50]. The new composition $Cs_3Bi_2Br_9$:Eu was further applied to detect Cu^{2+} through an electron transfer mechanism.

Considering the adverse effects of mercury ions, the sensing of mercury ions is highly important in biomedical applications. Lu et al. designed a sensor probe by using $CH_3NH_3PbBr_3$ perovskite quantum dots (QDs) for rapid visual detection of ultratrace mercury ions (Hg^{2+}) through an ion exchange mechanism with a detection limit of 0.124 nM (24.87 ppt) and high selectivity and sensitivity [51].

Wang et al. designed a novel fluorescent sensor based on $CsPbBr_3$ perovskite QDs encapsulated with the in situ hydrolysis of 3-aminopropyltriethoxysilane (APTES) for the selective detection of tetracycline in the polar solvent ethanol, with a lower detection limit of 76 nM [52]. The mechanism of fluorescence quenching takes place mainly through electron transfer between the TC and perovskite QDs. The TRPL decay graph shows that the average fluorescence lifetime of APTES@IPQDs decreased from 92.78 to 61.38 ns with the addition of 15 μM TC. The shortened fluorescence lifetime indicates that the sensing process involves a dynamic quenching mechanism. In addition, the sensor probe was successfully applied to the detection of TC in soil samples.

Later, methylammonium lead halide perovskite quantum dots (MAPB-QDs) were successfully applied for the detection of polar organochlorine pesticides (OCPs) on the basis of the blueshift of the MAPB-QD fluorescence spectra in the presence of the analyte, with a detection limit of 0.02 μM [53]. The detection process was further successfully applied to several real samples, such as apples and grapes.

7 Perovskite Quantum Dots as an Electrochemical Sensor

An electrochemical sensing platform is another method that delivers a convenient and low-cost detection method to detect many analytes. It has been widely applied in biomedical and environmental applications. In general, an electrochemical sensing system comprises three essential parts. A receptor binds with the respective analyte, the specific sample, and then the transducer part to convert the changes into some readable electrical signal. To report signals, many variables, such as voltage, electrochemical impedance, power output, and current, are used [54]. During detection, anodic stripping voltammetry (ASV) is commonly used. First, the analyte is deposited on the electrode where oxidation takes place [55].

Chen et al. designed the photoelectrochemical (PEC) sensor probe $CsPbBr_{1.5}I_{1.5}$ PQD-immobilized three-dimensional (3D) TiO_2 inverse opal photonic crystals (IOPCs) for the detection of dopamine (DA) [56]. The sensor probe can detect dopamine with a lower detection limit of approximately 0.012 μM. The composite films are excited by 500 W xenon light illumination. The band gaps of the $CsPbBr_{1.5}I_{1.5}$ QDs and TiO_2 are approximately 2.1 and 3.2 eV, respectively. The conduction band edge of the $CsPbBr_{1.5}I_{1.5}$ QDs is greater than that of TiO_2. Owing to the matching of the conduction bands of the TiO_2 and $CsPbBr_{1.5}I_{1.5}$ QDs under visible light illumination, the photogenerated electrons from the conduction band of the perovskite QDs transferred to the FTO through the TiO_2 layer, which suppressed direct charge recombination in the perovskite QDs. On the other hand, the holes in the $CsPbBr_{1.5}I_{1.5}$ QDs are transported from the valence band to the surface of the perovskite QDs (Fig. 11). Moreover, for the practical applicability of the sensor probe, a $TiO_2/CsPbBr_{1.5}I_{1.5}$/Nafion composite film was tested in human serum, with analytical recoveries in the range of 99–102%, which proves the biomedical application of the sensor probe.

Zhu et al. designed a photoelectrochemical (PEC) sensor based on $CsPbBr_3$/ZnO for the detection of dihydronicotinamide adenine dinucleotide (NADH) with a lower detection limit of 0.010 μM [57]. The band gap of ZnO is 3.2 eV, and the band gap of the $CsPbBr_3$ QDs is 2.38 eV. The conduction band edge of perovskite QDs is greater than the conduction band edge of ZnO. Under visible light illumination, the photogenerated electrons are transferred from the conduction band of the perovskite QDs to the FTO through the ZnO, resulting in direct charge recombination (Fig. 12). Furthermore, owing to the energy difference in the valence band between the $CsPbBr_3$ QDs and ZnO, the photogenerated holes are transferred to the surface of the perovskite QDs.

Chan et al. synthesized all perovskite nanowire $CsPbBr_3$ (PVSK NW) and $CsPb_{1-x}Sn_xBr_3$ (PVSK-Sn NW) and $CsPb_{1-x}Ge_xBr_3$ (PVSK-Ge NW) nanowires [58]. The PVSK-Sn NW was able to detect H_2O_2 with a lower detection limit of 0.12 μM. Cyclic voltammetry and electrochemical impedance analysis were used to detect the analyte H_2O_2. The plausible mechanism for the sensing process is as follows:

$$\mathrm{PVSK\,NW} + \mathrm{H_2O_2} + \mathrm{e^-} \rightarrow \mathrm{PVSK\,NW} + \mathrm{OH\,ad} + \mathrm{OH^-}$$

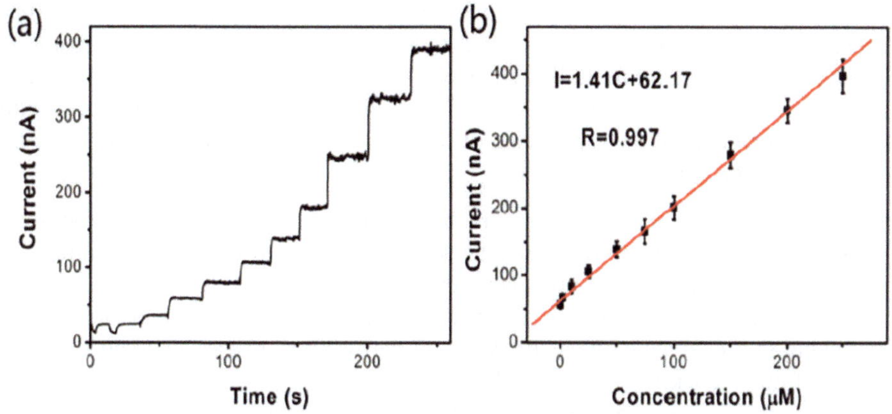

Fig. 11 (**a**) Photocurrent response of the $TiO_2/CsPbBr_{1.5}I_{1.5}$/Nafion electrode under 600 nm excitation with different concentrations of DA. (**b**) Calibration curve of the detection process

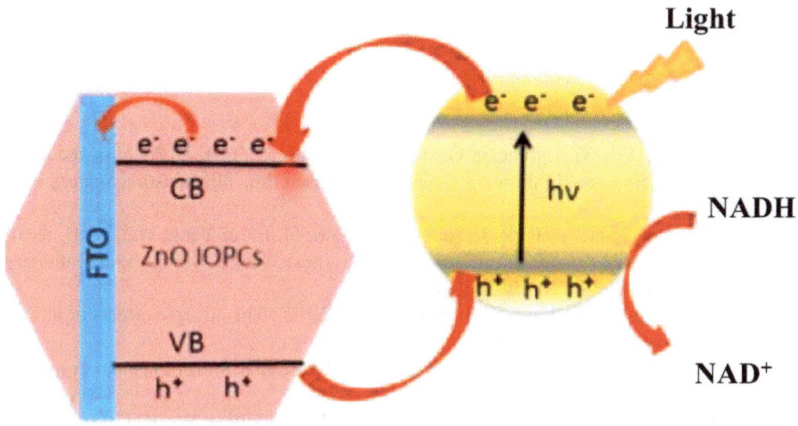

Fig. 12 Plausible mechanism of NADH detection by $CsPbBr_3$/ZnO perovskite quantum dot

$$\text{PVSK NW} + \text{OH ad} + e^- \rightarrow OH^-$$

$$2OH^- + 2H^+ \rightarrow H_2O$$

8 Conclusion

While addressing the importance of a sensor device, it is necessary to develop low-cost, high-sensitivity, and easy-to-operate system sensors using the most advantageous semiconductor materials, perovskite quantum dots, and thereafter, the expansion of the method towards commercialization. The proper protocol for high-accuracy sensors is not only beneficial to our environment but also an essential requirement in human life. In this chapter, we summarize various methods or techniques used during the sensing process for various analytes by using perovskite quantum dot materials. On the basis of the above-mentioned sensing utilities, the following points must be focused on:

- Since perovskite quantum dots are highly luminescent materials, a focus on fluorescent-based perovskite analytes should be given to sensing applications.
- Some sensor devices are influenced by elevated temperatures. Much optimization is necessary to achieve a stable sensory response at room temperature.
- Owing to the adverse effects of Pb-based perovskites, research should focus on lead-free perovskites as sensory devices.
- Other methods of perovskite sensing, such as electrical sensors and mass-sensitive detection with biosensors, have been investigated.

Acknowledgement The authors would like to acknowledge Tezpur University, Assam, India.

References

1. Xu, X., Zhou, J., Xin, Y., Lubineau, G., Ma, Q., & Jiang, L. (2017). Alcohol recognition by flexible, transparent, and highly sensitive graphene-based thin-film sensors. *Scientific Reports, 7*(1), 1–10.
2. Shuai, J., Kim, S., Ryu, H., Park, J., Lee, C. K., Kim, G. B., & Yang, W. (2018). Health risk assessment of volatile organic compounds exposure near Daegu dyeing industrial complex in South Korea. *BMC Public Health, 18*(1), 1–13.
3. Genc, S., Zadeoglulari, Z., Fuss, S. H., & Genc, K. (2012). The adverse effects of air pollution on the nervous system. *Journal of Toxicology, 2012*.
4. Ahn, K. (2014). The role of air pollutants in atopic dermatitis. *Journal of Allergy and Clinical Immunology, 134*(5), 993–999.
5. Otto, D. A., Hudnell, H. K., House, D. E., Mølhave, L., & Counts, W. (1992). Effects of exposure to a volatile organic mixture. Behavioral assessment. *Archives of Environmental Health, 47*, 23–30.
6. Mayes, A. G., Blyth, J., Kyröläinen-Reay, M., Millington, R. B., & Lowe, C. R. (1999). A holographic alcohol sensor. *Analytical Chemistry, 71*(16), 3390–3396.
7. Boroujerdi, R., Abdelkader, A., & Paul, R. (2020). State of the art in alcohol sensing with 2D materials. *Nano-Micro Letters, 12*(1), 1–33.
8. Paasma, R., Hovda, K. E., Tikkerberi, A., & Jacobsen, D. (2007). Methanol mass poisoning in Estonia: Outbreak in 154 patients. *Clinical Toxicology, 45*, 152–157.
9. García-Cañas, V., Simó, C., Herrero, M., Ibáñez, E., & Cifuentes, A. (2012). Present and future challenges in food analysis: foodomics. *Analytical Chemistry, 84*(23), 10150–10159.
10. Vickers, N. J. (2017). Animal communication: When I'm calling you, will you answer too? *Current Biology, 27*(14), R713–R715.
11. Martinez, A. W., Phillips, S. T., Whitesides, G. M., & Carrilho, E. (2010). Diagnostics for the developing world: microfluidic paper-based analytical devices. *Analytical Chemistry, 82*(1), 3–10.
12. Escarpa, A. (2014). Lights and shadows on food microfluidics. *Lab on a Chip, 14*(17), 3213–3224.

13. Ali, E. M. A., & Edwards, H. G. M. (1969). The detection of flunitrazepam in beverages using portable Raman spectroscopy. *Drug Testing and Analysis, 9*(2), 256–259. (2017).
14. Wei, W. Y., & White, I. M. (2013). Inkjet-printed paper-based SERS dipsticks and swabs for trace chemical detection. *Analyst, 138*(4), 1020–1025.
15. Mostowtt, T., & McCord, B. (2017). Surface-enhanced Raman spectroscopy (SERS) as a method for the toxicological analysis of synthetic cannabinoids. *Talanta, 164*, 396–402.
16. Tığ, G. A. (2017). Highly sensitive amperometric biosensor for determination of NADH and ethanol based on au-ag nanoparticles/poly (L-cysteine)/reduced graphene oxide nanocomposite. *Talanta, 175*, 382–389.
17. Zakrzewska, K., & Radecka, M. (2017). TiO 2-based nanomaterials for gas sensing—Influence of anatase and rutile contributions. *Nanoscale Research Letters, 12*(1), 1–8.
18. Bhatia, S., Verma, N., & Bedi, R. K. (2017). Ethanol gas sensor based upon ZnO nanoparticles prepared by different techniques. *Results in Physics, 7*, 801–806.
19. Wang, Y., Ding, G., Mao, J. Y., Zhou, Y., & Han, S. T. (2020). Recent advances in synthesis and application of perovskite quantum dot based composites for photonics, electronics and sensors. *Science and Technology of Advanced Materials, 21*(1), 278–302.
20. Chen, D., & Chen, X. (2019). Luminescent perovskite quantum dots: Synthesis, microstructures, optical properties and applications. *Journal of Materials Chemistry C, 7*(6), 1413–1446.
21. Kojima, A., Teshima, K., Shirai, Y., & Miyasaka, T. (2009). Organometal halide perovskites as visible-light sensitizers for photovoltaic cells. *Journal of the American Chemical Society, 131*(17), 6050–6051.
22. Park, N. G. (2015). Perovskite solar cells: An emerging photovoltaic technology. *Materials Today, 18*(2), 65–72.
23. Niu, G., Guo, X., & Wang, L. (2015). Review of recent progress in chemical stability of perovskite solar cells. *Journal of Materials Chemistry A, 3*(17), 8970–8980.
24. Weber, D. (1978). CH3NH3PbX3, ein Pb (II)-system mit kubischer perowskitstruktur/ CH3NH3PbX3, a Pb (II)-system with cubic perovskite structure. *Zeitschrift für Naturforschung B, 33*(12), 1443–1445.
25. Kitazawa, N., Yamamoto, T., Watanabe, Y., & Nakamura, Y. (1999). Stability of self-assembled organic–inorganic layered perovskite. *MRS Online Proceedings Library (OPL), 576*, 165.
26. Babin, V., Fabeni, P., Nikl, M., Pazzi, G. P., Sildos, I., Zazubovich, N., & Zazubovich, S. (1999). Polarized luminescence of CsPbBr3 nanocrystals (quantum dots) in CsBr: Pb single crystal. *Chemical Physics Letters, 314*(1–2), 31–36.
27. Protesescu, L., Yakunin, S., Bodnarchuk, M. I., Krieg, F., Caputo, R., Hendon, C. H., et al. (2015). Nanocrystals of cesium lead halide perovskites (CsPbX3, X= Cl, Br, and I): Novel optoelectronic materials showing bright emission with wide color gamut. *Nano Letters, 15*(6), 3692–3696.
28. Li, X., Wu, Y., Zhang, S., Cai, B., Gu, Y., Song, J., & Zeng, H. (2016). CsPbX3 quantum dots for lighting and displays: Room-temperature synthesis, photoluminescence superiorities, underlying origins and white light-emitting diodes. *Advanced Functional Materials, 26*(15), 2435–2445.
29. Jellicoe, T. C., Richter, J. M., Glass, H. F., Tabachnyk, M., Brady, R., Dutton, S. E., et al. (2016). Synthesis and optical properties of lead-free cesium tin halide perovskite nanocrystals. *Journal of the American Chemical Society, 138*(9), 2941–2944.
30. Zhang, J., Yang, Y., Deng, H., Farooq, U., Yang, X., Khan, J., et al. (2017). High quantum yield blue emission from lead-free inorganic antimony halide perovskite colloidal quantum dots. *ACS Nano, 11*(9), 9294–9302.
31. Zhou, Q., Bai, Z., Lu, W. G., Wang, Y., Zou, B., & Zhong, H. (2016). In situ fabrication of halide perovskite nanocrystal-embedded polymer composite films with enhanced photoluminescence for display backlights. *Advanced Materials, 28*(41), 9163–9168.
32. Chen, Z., Gu, Z. G., Fu, W. Q., Wang, F., & Zhang, J. (2016). A confined fabrication of perovskite quantum dots in oriented MOF thin film. *ACS Applied Materials & Interfaces, 8*(42), 28737–28742.

33. Zhang, F., Shi, Z., Li, S., Ma, Z., Li, Y., Wang, L., et al. (2019). Synergetic effect of the surfactant and silica coating on the enhanced emission and stability of perovskite quantum dots for anticounterfeiting. *ACS Applied Materials & Interfaces, 11*(31), 28013–28022.
34. Yan, D., Shi, T., Zang, Z., Zhou, T., Liu, Z., Zhang, Z., et al. (2019). Ultrastable CsPbBr3 perovskite quantum dot and their enhanced amplified spontaneous emission by surface ligand modification. *Small, 15*(23), 1901173.
35. Wang, Z., Fu, R., Li, F., Xie, H., He, P., Sha, Q., et al. (2021). One-step polymeric melt encapsulation method to prepare CsPbBr3 perovskite quantum dots/polymethyl methacrylate composite with high performance. *Advanced Functional Materials, 31*(22), 2010009.
36. Xie, Z., Li, X., Li, R., Lu, S., Zheng, W., Tu, D., et al. (2020). In situ confined growth of ultrasmall perovskite quantum dots in metal–organic frameworks and their quantum confinement effect. *Nanoscale, 12*(32), 17113–17120.
37. Han, X., Zhang, G., Li, B., Yang, C., Guo, W., Bai, X., et al. (2020). Blinking mechanisms and intrinsic quantum-confined stark effect in single methylammonium lead bromide perovskite quantum dots. *Small, 16*(51), 2005435.
38. Liu, H., & Bansal, S. (2023). Metal halide perovskite nanostructures and quantum dots for photocatalytic CO2 reduction: Prospects and challenges. *Materials Today, Energy, 32*, 101230.
39. Thungon, P. D., Kakoti, A., Ngashangva, L., & Goswami, P. (2017). Advances in developing rapid, reliable, and portable detection systems for alcohol. *Biosensors and Bioelectronics, 97*, 83–99.
40. Zhu, Z., Sun, Q., Zhang, Z., Dai, J., Xing, G., Li, S., Huang, X., & Huang, W. (2018). Metal halide perovskites: Stability and sensing-ability. *Journal of Materials Chemistry C, 6*(38), 10121–10137.
41. Halali, V. V., Sanjayan, C., Suvina, V., Sakar, M., & Balakrishna, R. G. (2020). Perovskite nanomaterials as optical and electrochemical sensors. *Inorganic Chemistry Frontiers, 7*(14), 2702–2725.
42. Zhu, Z., Sun, Q., Zhang, Z., Dai, J., Xing, G., Li, S., et al. (2018). Metal halide perovskites: Stability and sensing-ability. *Journal of Materials Chemistry C, 6*(38), 10121–10137.
43. Geies, A., & Abdelazim, M. (2021). Self-colorimetric determination of bio-ethanol using permanganate in fermentation samples. *Open Access Library Journal, 8*(8), 1–8.
44. De Giorgi, M. L., Milanese, S., Klini, A., & Anni, M. (2021). Environment-induced reversible modulation of optical and electronic properties of lead halide perovskites and possible applications to sensor development: A review. *Molecules, 26*(3), 705.
45. Park, B., Kang, S. M., Lee, G. W., Kwak, C. H., Rethinasabapathy, M., & Huh, Y. S. (2019). Fabrication of CsPbBr3 perovskite quantum dots/cellulose-based colorimetric sensor: Dual-responsive on-site detection of chloride and iodide ions. *Industrial & Engineering Chemistry Research, 59*(2), 793–801.
46. Huangfu, C., & Feng, L. (2021). High-performance fluorescent sensor based on CsPbBr3 quantum dots for rapid analysis of total polar materials in edible oils. *Sensors and Actuators B: Chemical, 344*, 130193.
47. Wang, Y., Zhu, Y., Huang, J., Cai, J., Zhu, J., Yang, X., et al. (2016). CsPbBr3 perovskite quantum dots-based monolithic electrospun fibre membrane as an ultrastable and ultrasensitive fluorescent sensor in aqueous medium. *The Journal of Physical Chemistry Letters, 7*(21), 4253–4258.
48. Huang, S., Guo, M., Tan, J., Geng, Y., Wu, J., Tang, Y., et al. (2018). Novel fluorescence sensor based on all-inorganic perovskite quantum dots coated with molecularly imprinted polymers for highly selective and sensitive detection of omethoate. *ACS Applied Materials & Interfaces, 10*(45), 39056–39063.
49. Sheng, X., Liu, Y., Wang, Y., Li, Y., Wang, X., Wang, X., et al. (2017). Cesium lead halide perovskite quantum dots as a photoluminescence probe for metal ions. *Advanced Materials, 29*(37), 1700150.

50. Ding, N., Zhou, D., Pan, G., Xu, W., Chen, X., Li, D., et al. (2019). Europium-doped lead-free Cs3Bi2Br9 perovskite quantum dots and ultrasensitive Cu2+ detection. *ACS Sustainable Chemistry & Engineering, 7*(9), 8397–8404.
51. Lu, L. Q., Tan, T., Tian, X. K., Li, Y., & Deng, P. (2017). Visual and sensitive fluorescent sensing for ultratrace mercury ions by perovskite quantum dots. *Analytica Chimica Acta, 986*, 109–114.
52. Wang, T., Wei, X., Zong, Y., Zhang, S., & Guan, W. (2020). An efficient and stable fluorescent sensor based on APTES-functionalized CsPbBr 3 perovskite quantum dots for ultrasensitive tetracycline detection in ethanol. *Journal of Materials Chemistry C, 8*(35), 12196–12203.
53. Yang, Y., Han, A., Hao, S., Li, X., Luo, X., Fang, G., et al. (2020). Fluorescent methylammonium lead halide perovskite quantum dots as a sensing material for the detection of polar organochlorine pesticide residues. *Analyst, 145*(20), 6683–6690.
54. Baranwal, J., Barse, B., Gatto, G., Broncova, G., & Kumar, A. (2022). Electrochemical sensors and their applications: A review. *Chem, 10*(9), 363.
55. Guziejewski, D. (2019). Square-wave amplitude effect in cathodic and anodic stripping square-wave voltammetry. *Electroanalysis, 31*(2), 231–238.
56. Chen, X., Li, D., Pan, G., Zhou, D., Xu, W., Zhu, J., et al. (2018). All-inorganic perovskite quantum dot/TiO 2 inverse opal electrode platform: Stable and efficient photoelectrochemical sensing of dopamine under visible irradiation. *Nanoscale, 10*(22), 10505–10513.
57. Zhu, Y., Tong, X., Song, H., Wang, Y., Qiao, Z., Qiu, D., et al. (2018). CsPbBr 3 perovskite quantum dots/ZnO inverse opal electrodes: Photoelectrochemical sensing for dihydronicotinamide adenine dinucleotide under visible irradiation. *Dalton Transactions, 47*(30), 10057–10062.
58. Lu, L. Q., Ma, M. Y., Tan, T., Tian, X. K., Zhou, Z. X., Yang, C., & Li, Y. (2018). Novel dual ligands capped perovskite quantum dots for fluoride detection. *Sensors and Actuators B: Chemical, 270*, 291–297.
59. Zhu, R., Du, Z., Zhu, M., Liang, H., Wang, S., Zhou, Q., et al. (2023). Molecularly imprinted polymers embedded with double perovskite quantum dots: A ratiometric fluorescence sensor for visible and fluorescent determination of Rhein. *Chemical Engineering Journal, 468*, 143618.
60. Ahmed, S., Mohanta, D., & Dolui, S. K. (2023). CsPbBr 3 perovskite quantum dot decorated ZIF-8 MOF: A selective dual recognition fluorometric visual probe for 4-nitroaniline and rhodamine blue. *Analytical Methods, 15*(46), 6394–6403.
61. Liu, L., Peng, M., Xu, K., Xia, H., Peng, X., Peng, L., & Zhang, J. Z. (2023). Molecularly imprinted fluorescence assay based on lead halide perovskite quantum dots for determination of benzo (a) pyrene. *Microchimica Acta, 190*(10), 380.
62. Gao, D., Zhang, A., Lyu, B., & Ma, J. (2024). Visual and rapid fluorescence sensing for hexavalent chromium by hydroxypropyl chitosan passivated bismuth-based perovskite quantum dots. *Microchimica Acta, 191*(4), 219.
63. Sanjayan, C. G., Ravikumar, C. H., & Balakrishna, R. G. (2023). Perovskite QD based paper microfluidic device for simultaneous detection of lung cancer biomarkers–carcinoembryonic antigen and neuron specific enolase. *Chemical Engineering Journal, 464*, 142581.

Recent Advances in Boron Nitride Quantum Dots in Biomedical Applications

Debika Gogoi, Devipriya Gogoi, and Manash R. Das

1 Introduction

Zero-dimensional (0D) fluorescent quantum dots (QDs) are defined as tiny nanomaterials with diameters in the nanoscale region, specifically those with radii smaller than the Bohr excitonic radius. These tiny particles, which typically range from 2 to 10 nm in diameter, exhibit thermal, optical, and electrical characteristics that are dissimilar from those of bulk materials owing to their very small scale [1–3]. This size-dependent property makes quantum dots highly valuable in various applications. QDs were first synthesized and studied in 1981 by A.I. Ekimov and A.A. Onushchenko, and since then, various 0D QDs have been widely studied by different researchers [4].

Elemental boron (B) has three valence electrons and lies between metals and nonmetallic elements in the periodic table. Boron (B) has multiple bonding characteristics because of the electron deficiency in both two-center, two-electron (2c-2e) and three-center, two-electron (3c-3e) B-B bonds [5, 6]. These atypical bonding conditions propel the elemental boron to exhibit at least 16 polymorphs that possess strength, firmness, durability, and stability, which expands the potential field in advanced high-temperature systems [5, 7]. In contrast, compared with boron materials, nanostructured boron materials are more likely to have extensive and beneficial applications. To date, various nanostructured boron materials, such as zero-dimensional (0D), one-dimensional (1D), and two-dimensional (2D) boron quantum dots (BQDs), boron nanowires, and boron nanosheets, respectively, have been widely studied (experimentally and theoretically) by researchers.

D. Gogoi · D. Gogoi · M. R. Das (✉)
Coal, Energy and Materials Sciences Division, CSIR-North East Institute of Science and Technology, Jorhat, Assam, India
e-mail: manashrdas.neist@csir.res.in

S. P. Kabekkodu et al. (eds.), *Functional Quantum Dots*, Lecture Notes in Nanoscale Science and Technology 40,
https://doi.org/10.1007/978-3-031-98219-4_10

For the past few decades, considerable research progress has been made on QDs in the field of nanomaterials. However, over the past few years, boron-based nanostructures have attracted increased interest in research areas because of their excellent properties and biocompatibility [8–10]. Boron nitride QDs (BNQDs) are an emerging class of nanomaterials that have undergone rapid growth in various fields of applied science, such as bioimaging, sensing, near-infrared (NIR) light-induced photothermal therapy drug delivery, and optoelectronics, owing to their exceptional stability, low toxicity, and tunable photoluminescence [11–14].

2 Synthetic Approaches for Boron Nitride Quantum Dots

To date, various synthetic approaches have been established for the preparation of boron nitride quantum dots (BNQDs). These developed synthesis techniques can be classified into two types: top-down and bottom-up approaches. The enactment of these two synthesis routes depends on the purity and size of the final product, the total time requirement for synthesis and its affordability. The top-down methods usually depend on the liquid exfoliation of bulk boron nitride (BN) using a suitable solvent. Generally, bulk boron nitride powder is exfoliated in the presence of a solvent, which generates boron nitride nanosheets (BNNSs) with layer-by-layer sheets in the transition state, followed by breakdown to produce BNQDs. The bottom-up methods involve the interactions of boron precursors at the molecular level. Typically, solvent-assisted or liquid-phase approaches are utilized for the production of BNQDs. Moreover, top-down and liquid-phase approaches can be combined to enhance the synthesis of fluorescent BNQDs from bulk BN.

2.1 Top-Down Methods

2.1.1 Liquid Exfoliation Method

The liquid-phase exfoliation (LPE) method is a simple, safe, and environmentally friendly approach for synthesizing BNQDs. The LPE technique is usually used with liquid solvents that have adequate surface energy, such as DMF, ethanol, and NMP. In this method, bulk hexagonal boron nitride (h-BN) with a large size is exfoliated and then itemized to prepare BNQDs, and a subsequent separation technique is then employed via the centrifuge technique [3, 9, 15]. Li et al. prepared BNQDs via the LPE method, in which micrometer (μm)-sized BNNSs were fragmented via a solvothermal process owing to its simplicity and ability to be implemented on a large scale [16]. High-grade BNQDs are consistently synthesized by changing various parameters, such as the exfoliation time, temperature, and blending factor. Intense blue luminescence is observed in the prepared BNQDs, which is an outcome of size–edge defect interactions [16]. Kumar et al. synthesized BNQDs from bulk h-BN with an average thickness of 12 (unit) atomic h-BN layers via a one-step technique involving high-frequency sonication, which is associated with the LPE technique [17]. Bulk BN was exfoliated into BNNSs via sonication with phosphoric acid (H_3PO_4). The acquired solution was repeatedly washed with distilled water to remove any residual acid after the completion of the sonication process. The suspension was washed repeatedly until a pH of ~7 was reached. Then, the small and large BN nanoparticles were separated via centrifugation at different speeds. Typically, small BN nanoparticles float in a solvent, whereas large BN nanosheets precipitate at the bottom of the container. Further investigation revealed that the BN nanoparticles that settled at the bottom of the container comprised a few thin layers or nanosheets, whereas the small BN nanoparticles that floated on the water were BN quantum dots [17]. The fragmentation of BNNSs via high-power sonication of bulk h-BN occurred as a result of consecutive changes in pressure, which were mechanically generated throughout the sonication process, leading to the formation of QDs [17]. At the initial stage, the weak van der Waal force holds these thick layers of BNs together; thus, mechanical pressure is applied along their side edges, which leads to the breakdown of BN powder into smaller and thinner BNNSs [18]. The intercalation of h-BN by H_3PO_4 facilitates the simple delamination of thicker BNNSs into thinner ones [19]. The thinner BNNSs finally breakdown in the final step and fragment into smaller BNNSs, as they are not capable of enduring the applied mechanical pressure, resulting in the formation of BNQDs. A plausible mechanism of BNQD formation is depicted in Fig. 1 [17].

Additionally, BNQDs were synthesized by the fragmentation of h-BNNs by Stengl et al. exposed BNNSs to highly intense ultrasonication [20]. They utilized ultrasonication to proficiently delaminate h-BN into nanosheets comprising a few layers, which is a very good precursor for the design of pure BNQDs. First, bulk h-BN was prepared via a chemical technique in which urea and boric acid were used as precursors comprising a structure with a number of layers. The number of BN layers decreased as the amount of urea increased. Afterwards, a highly intense cavitation field was employed in a pressurized gas reactor for the conversion of bulk h-BN to delaminated h-BN. The h-BN was first dispersed in solvents such as isopropanol (IPA), ethylene glycol, distilled water or ethylene glycol, after which a strong cavitation field was applied to the pressurized gas reactor for approximately 5 min. BN powder was typically treated with H_2SO_4 to prepare exfoliated BNNSs, which was followed by the addition of $KMnO_4$. The mixture was subsequently heated at a constant temperature of 40 °C with continuous stirring for 6 h. A pink suspension was formed, to which ~200 mL of 30% H_2O_2 was added while the mixture was in an ice bath. The pink suspension was instantly transformed into a white suspension. This suspension was centrifuged and washed until a pH of 7.0 or slightly lower was obtained. The obtained exfoliated h-BNNSs were then refluxed with ethylene glycol for 48 h at atmospheric pressure to yield BNQDs. The resulting dispersion was removed via a Pragopor membrane filter, and a clear solution was obtained [20]. Thangasamy et al. established a BNQD synthesis procedure using a supercritical fluid (SCF) process for the rapid and effortless production of blue light-emitting fluorescent BNQDs from BN layers via in situ exfoliation followed by

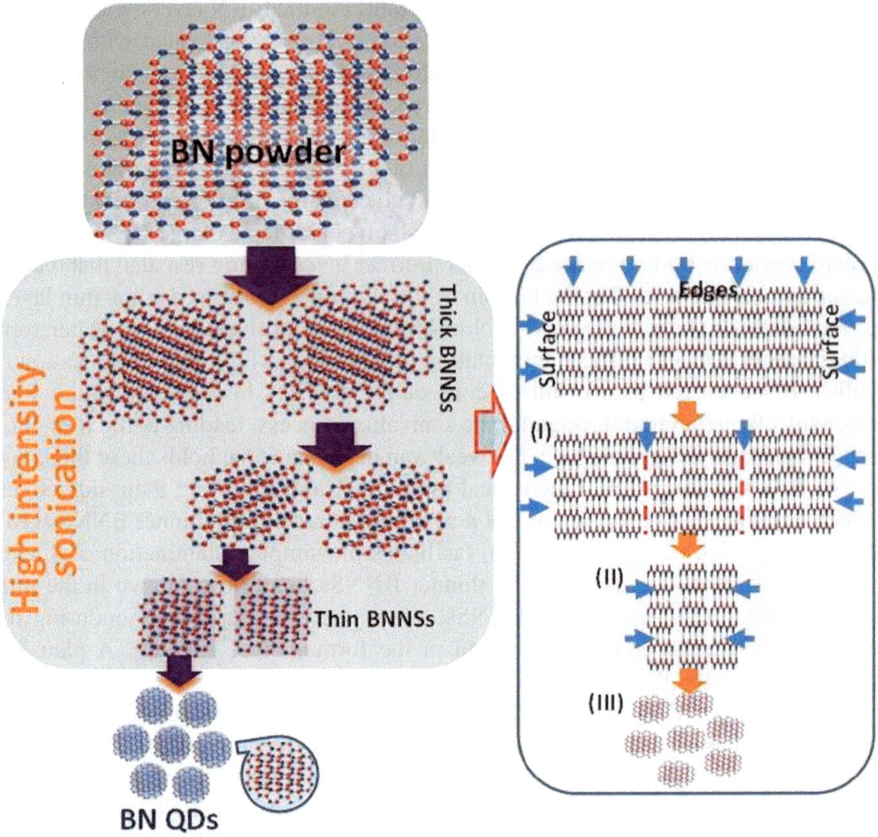

Fig. 1 Proposed mechanism for the conversion of BNQDs from BN nanoparticles using high intense sonication technique. (Copyright (2016) J. Alloys Compds [17])

disintegration [21]. The formation of few-layered BNQDs was verified by atomic force microscopy (AFM) and microscopic analysis. SCF is a distinct and homogenous reaction medium that is prepared by levitating the temperature and pressure of the solvent to the point after which separate liquid and gas phases tend to exist. Moreover, SCFs combine vapor and liquid features, for example, high diffusion coefficients, low interfacial tension, and high surface wetting [9, 22]. The production of various metal oxide nanoparticles, in addition to the functionalization and delamination of graphene nanosheets, can be achieved from these special properties. The formation of BNQDs was identified by the fragmentation of the exfoliated monolayered or few-layered BN nanosheets after the exfoliation of the BN nanosheets in the SCF. Typically, bulk h-BN powder is dispersed in a solvent (DMF), and then, this reaction mixture is treated ultrasonically and then transferred to a sealed autoclave reactor, which is placed in a preheated (450 °C) vertical tubular furnace. The reactor was then immediately drenched with cold water. The subsequent solution was centrifuged and filtered via a membrane filter with a micropore size of 0.22 m. The resulting mildly yellow-colored filtrate containing the BNQD dispersion was collected and characterized thoroughly via high-resolution transmission electron microscopy (HRTEM), and an average particle size of 5 nm was observed [21]. In conclusion, the LPE method is a powerful and effective technique for synthesizing BNQDs because of its cost-effectiveness, simplicity, and feasibility. However, this LPE approach also has several disadvantages, such as the size of the produced BNQDs, which is mostly governed by the solvent polarity. The high polarity of the solvent used facilitates the weakening of the interaction force between two adjacent layers of h-BN layers, which stimulates the exfoliation of the bulk h-BN to generate h-BNNSs.

2.1.2 Ion Intercalation Method

In the ion intercalation method, ions are introduced into the crystal lattice of BN to adjust the interlayer distance and generate exfoliation. The benefits of the synthesized BNQDs obtained via this process include their very small size and thin structure. Li, Na, and K are the primary intercalation ions used in this method. These ions can be inserted into the interlayers to facilitate exfoliation, which is then combined with other techniques to synthesize BNQDs. This is possibly due to the significant distance and weak van der Waals forces of attraction between h-BN layers. [3, 9] The same procedure was employed by Liu et al. to prepare BNQDs from NaOH-treated bulk BN powder using ethanol as a solvent [23]. Bulk h-BN was dispersed in ethanol, followed by the addition of NaOH to this dispersion, which was then stirred for 10 min. Subsequently, the solution mixture was heated for 24 h at 180 °C via an autoclave. After that, the mixture was cooled to room temperature, followed by freeze-drying, which yielded yellow BNQDs. The role of the ethanol solvent was to increase the contact interface between h-BN and the intercalating agent, resulting in increased solubility of bulk h-BN. The use of NaOH may disintegrate the edges of h-BN sheets under solvothermal conditions, which assists the intercalation of Na^+ and OH^- ions between the h-BN layers. Moreover, Vander Waal's force of interaction between the layers is weakened by high pressure and temperature, which improves the interactions of h-BN with NaOH. The exfoliation of layers in bulk h-BN eventually results in the formation of BNNSs and BNQDs, with the insertion of NaOH between the BN layers. The crystallographical and morphological aspects of the synthesized BNQDs were studied via HRTEM, which revealed that their average diameter was 5.1 nm [23]. Similarly, Lin et al. synthesized BNQDs via this method by intercalating potassium atoms between the layers of h-BN [24]. First, K atoms are subjected to bulk h-BN ad because of the weak van der Waal force of attraction between h-BN layers, and the intercalation process becomes easy. This subsequently forms K-h-BN, which is exposed to air for a short time span, which aids in the deintercalation process. In the final step, K-h-BN was allowed to interact with EtOH-H_2O, which was mediated ultrasonically. The h-BN layers disintegrated and exfoliated during the reaction of K-h-BN with EtOH-H_2O and in contact with air [24]. However, the reaction conditions and process parameters allowed the intercalation of K atoms into the gaps between the layers of h-BN, which also resulted in changes in color. The temperature applied in this intercalation process was 190–200 °C, which is markedly lower than that of the formerly reported methods for the synthesis of K-h-BN (300 °C) [9, 25]. Yellow-colored BNQDs were finally obtained, and even after 5 months of synthesis, there was no precipitation. HRTEM analysis revealed that these stacked BNQDs have an interlayer distance of 5–8 Å, which is remarkably greater than that of h-BN (002). This indicated the formation of monolayered BNQDs because of the exfoliation and disintegration of h-BN flakes [24]. Peng et al. utilized a combination of intercalation and ultrasonic exfoliation methods to prepare BNNSs, which were further used for the fabrication of BNQDs [26]. First, 5% N_2H_4 was added to the bulk BN powder to make a dilute mixture, which was then transferred to a stainless-steel autoclave assisted with Teflon and heated to 150 °C for 24 h. After the mixture cooled to room temperature, the mixture was centrifuged, and the precipitate containing BN was collected for later use and characterization. The processed BN was added to a 55% ethanol–water mixture and ultrasonicated for 8 h. Then, the resulting milky white BNNS solution was heated at 80 °C in a vacuum oven to completely remove the ethanol, and the remaining BNNS suspension was collected. This suspension was treated with KOH to achieve a pH of ~8, and then, the BNNS suspension was heated at a temperature of 200 °C for approximately 12 h in a Teflon-lined autoclave. The resulting mixture was filtered through a microporous membrane (0.22 μm) to remove the larger tracts, and the final solution contained BNQDs that displayed intense blue fluorescence. N_2H_4 can be efficiently inserted into the interlayers of BN powder via the above synthesis protocol. A portion of the original BN was converted to thin bits owing to the formation of unstable $N_2H_5^+$ from the oxidation of N_2H_4, which can easily cease to form N_2, H_2, and NH_3. The average size of the as-synthesized BNQDs was 1.0–2.6 nm [26]. Generally, the ion intercalation procedure generates monolayered BNQDs. However, BNQDs prepared via this method also have several disadvantages, such as low quantum yield and large average particle size, compared with those of other synthesis protocols. Once the intercalating ions are inserted into the h-BN interlayers, their size and properties do not change during the subsequent exfoliation and rupture process. These ions further weaken the force of interaction between neighboring h-BN layers and stimulate the segmentation and exfoliation of h-BNNSs, which enhances the yield of BNQDs synthesized via the ion intercalation technique. Larger ions such as CO_3^{2-}, NH_4^+, and SO_4^{2-} may decompose to produce additional gases during subsequent rupture and exfoliation procedures [9, 27].

2.1.3 Laser Ablation Method

The laser ablation process is a top-down process that is employed to produce nanomaterials such as nanowires, QDs, and core-shell-type nanoparticles. In this method, nanoparticles are formed through nucleation growth followed by the development of laser-vaporized particles in a gaseous atmosphere. This method has the ability to generate high-grade quantum particles in the size range of <10 nm, which is assisted by the remarkably fast reduction of vapor [9, 28]. Nguyen et al. used a femtosecond laser ablation process to prepare BNQDs from bulk h-BN powder in solvents such as ethylenediamine, ethanol, and diethyl amine [29]. The h-BN powder was dispersed in the organic solution by sonication for 20 min. This mixture was then subjected to laser treatment along with stirring of the sample for 30 min. This process ultimately produced BNQDs, which were collected via the centrifugation technique. HRTEM analysis of the synthesized BNQDs revealed that the diameter of the BNQDs was in the range of 1.4–1.8 nm, and the d-spacing of the BNQDs was 0.21 nm [29]. Xu et al. established an upfront approach to synthesize BNQDs, which is a combination of femtosecond laser pulse ablation and ultrasonic-assisted exfoliation techniques [30]. Initially, bulk h-BN is fragmented into small nm-sized particles with the help of laser pulse ablation, followed by exfoliation into QDs with a few-layer thickness via ultrasonication. The combined method finally produces a pale-yellow mixture that comprises BNQDs. Additionally, N-methyl-2-pyrrolidone (NMP) was used as the solvent to reduce van der Waal's force of attraction between the h-BN layers. The equivalent solvent surface energy facilitates the exfoliation of several-layered nanoparticles into a few-layered or monolayered nanoparticles [30–32]. HRTEM analysis revealed that the obtained BNQDs were highly crystalline, well dispersed and had an average particle size of 2.1 nm. The surface morphology studies revealed that few-layered QDs were developed via the ultrasonic technique and that BN nanoparticles were produced via the laser pulse ablation method [30].

2.2 *Bottom-Up Methods*

2.2.1 Solvothermal/Hydrothermal Treatment

The solvothermal/hydrothermal synthesis technique is one of the most commonly used techniques for preparing various QDs. In this methodology, a microscopic molecule is used as a precursor, a reaction medium composed of water or organic solvents, and high pressure and temperature to yield the desired QDs [9, 32]. Chen et al. prepared BNQDs from h-BNNSs by exfoliation via a hydrothermal synthesis protocol [10]. Commercially available h-BN powder was dispersed in a solution composed of acetone, ethanol, and water, which was processed ultrasonically to generate h-BNNSs with good crystallinity and small lateral sizes. Afterward, the h-BNNSs were mixed with DMF, and the mixture was transferred to a stainless-steel Teflon-lined reactor for hydrothermal treatment at 180 °C to yield BNQDs. The resulting BNQDs had a homogeneous distribution of particles with an average size of 1.65 nm. Moreover, the BNQDs also exhibited intense blue fluorescence and nanosecond luminescent degradation upon exposure to UV-V irradiation. Figure 2 shows a graphical representation of the synthesis of h-BNNSs through the ultrasonic technique and BNQDs via the solvothermal technique and their respective FESEM micrographs along with digital pictures (inset) [10]. Initially, bulk boron

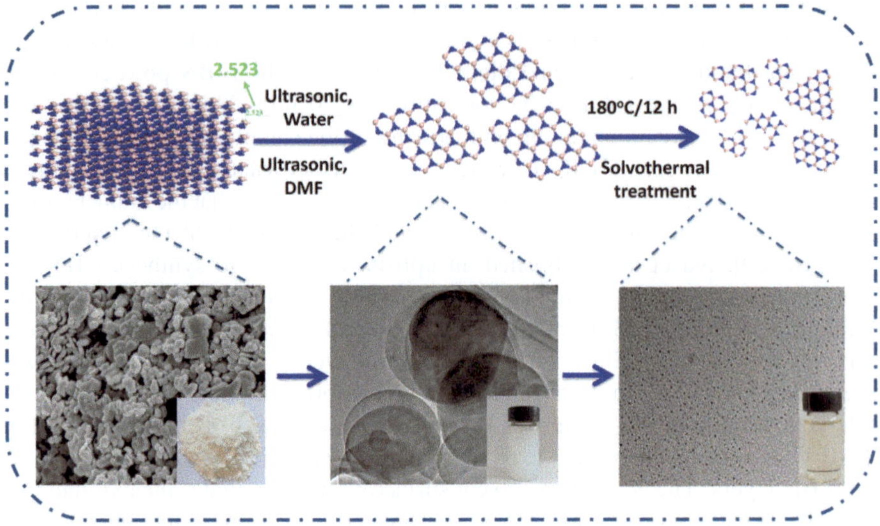

Fig. 2 The graphical representation of the synthesis of h-BNNSs and BNQDs with the respective FESEM images, digital pictures of the obtained materials. (Copyright (2021) Colloids Surf. A: Physicochem. Eng. Asp [10])

nitride was added to a solution of acetone, ethanol, or water, which was subjected to ultrasonication for 3 h. The layers of h-BN can be detached from each other in the course of this ultrasonication treatment with the incorporation of -OH groups between adjacent layers. The h-BN layers can be further exfoliated by ultrasonicating with dimethylformamide (DMF). The DMF solvent possesses a specific surface energy that helps reduce van der Waal's force of interaction between h-BN layers, subsequently assisting the fragmentation of h-BNNSs to form BNQDs [10]. Sert et al. applied a single-step hydrothermal synthesis procedure to synthesize BNQDs and studied their antifouling properties via the modification of polyethersolfone (PES) with BNQDs [33]. Urea and boric acid were used as precursors in this synthesis method. First, boric acid was added to a solution of ethanol and deionized water (molar ratio of 2:1). Simultaneously, urea was added to an ammonia solution, which was then added to the previous solution containing boric acid, ethanol, and water. The mixture was then stirred for 10 min, followed by hydrothermal treatment for 24 h at 200 °C, which yielded BNQDs. A composite membrane was fabricated via the modification of a PES membrane with BNQDs, which demonstrated enhanced permeability and antifouling properties. The nonsolvent-induced phase separation (NIPS) technique was used to fabricate this composite membrane. The crystal structure of the BNQDs was studied via XRD analysis, which confirmed the formation of BNQDs from the exfoliation of raw h-BN layers due to hydrothermal treatment. To study the morphology of the synthesized BNQDs, HRTEM analysis was used, which revealed that these BNQDs have a spherical shape with an average size of 3–4 nm [33]. Furthermore, the BNQDs that were prepared via a one-step solvothermal method using boric acid and melamine as boron and nitrogen sources, respectively, displayed strong blue fluorescence [34]. Boric acid and melamine were dissolved in distilled water and then transferred to a Teflon-lined autoclave, which was heated at 200 °C for 15 h. This process finally yielded BNQDs with a lateral size of ~3 nm that were highly crystalline in nature. Atomic force microscopy (AFM) analysis was used to further research the thickness and morphology of the BNQDs, which indicated that these BNQDs were monolayered. The prepared BNQDs were used for the sensing of Fe^{3+} via a fluorescence quenching process [34]. Liu et al. also synthesized BNQDs functionalized with hydroxyl and amine groups by combining boric acid and ammonia solutions as boron and nitrogen sources, respectively, following a single-pot hydrothermal procedure [35]. These functionalized BNQDs were also utilized in the detection of ferrous and ferric ions. The feasibility and benefits of the reported hydrothermal process and fluorescence quenching mechanism were also supported by density functional theory (DFT) simulations. The authors proposed an environmentally favorable and scalable way to synthesize BNQDs with high dispersibility, which can also be used to synthesize other 2-D QDs. Generally, the hydrothermal technique offers an effective way to achieve monolayered BNQDs in high yield. Choosing the nitrogen precursor is imperative during the whole procedure of the synthesis of BNQDs. Hence, major research emphasizes the quest to find environmentally friendly and inexpensive precursors for the fabrication of high-grade BNQDs [35]. Kamyabi et al. prepared BNQDs via a hydrothermal process in which boric acid and ammonia were used as the boron and nitrogen sources, respectively, and DMF was used as the solvent [36]. The synthesized BNQDs were utilized as a coreactant for cathodic ECL monitoring of chlorpyrifos. Ammonia was added to the solution containing boric acid and DMF, and then, this mixture was purged with nitrogen to eliminate the oxygen present. The reaction mixture was then subjected to ultrasonication for approximately 15 min, followed by hydrothermal treatment at 130 °C for 12 h. After the reaction mixture cooled to room temperature, the BNQDs were collected via centrifugation and filtration. The successful formation of the BNQDs was confirmed by HRTEM analysis, which also revealed that their average size was less than 10 nm. Similarly, the XRD pattern also showed a wide peak at 26.6 °C, which indicated the loss of crystallinity of BN during synthesis [36]. Furthermore, Chu et al. prepared a BNQD/Ti_3C_2MXene-based nanocomposite and utilized it in electrolytic nitrogen fixation [37]. The synthesis protocol involved the preparation of a dispersed suspension of BNNSs in DMF, which was then hydrothermally treated at 200 °C for 24 h. After the reaction mixture was cooled to room temperature, it was centrifuged and filtered, followed by freeze drying to finally obtain BNQDs with an average particle size of ~3.4 nm. The Al layers present in the Ti_3AlC_2 MAX phase were selectively etched out via a HCL/LiF mixture to yield $Ti_3C_2T_x$ MXene, which was then subjected to ultrasonication. This results in the formation of stacked $Ti_3C_2T_x$ nanosheets with few layers. Electrostatic interaction facilitates efficient anchoring of the BNQDs on $Ti_3C_2T_x$ nanosheets, which was carried out by simply mixing the BNQDs and $Ti_3C_2T_x$ nanosheets, yielding the required BNQD/$Ti_3C_2T_x$ heterostructure [37]. Wang et al. developed a feasible solvothermal approach to prepare BNQDs from

BNNSs and P-phenylenediamine [38]. First, the BNNSs were dispersed in 20 mL of ethanol, after which they were treated ultrasonically for 8 h. The resulting dispersion was subsequently centrifuged to remove any residual larger BN particles. A homogenous solution was then prepared by mixing the acquired supernatant, ethanol, and P-phenylenediamine, which were hydrothermally treated at 200 °C for 15 h [38]. Phenylenediamine acts as a passivator in the reaction, which suppresses the oxidation process and facilitates the formation of BNQDs. The synthesized BNQDs showed modular fluorescence properties, which thereby increased the amination process associated with surface states. The formation of BNQDs was examined via HRTEM, which revealed a fine size distribution of ~2.3–7.7 nm and an average diameter of ~4.9 nm [38].

2.2.2 Microwave Method

Compared with other synthetic methods, the microwave-assisted synthesis method is a very beneficial procedure in which a conventional source of energy is used. This method shortens the reaction time and enhances the yield, and the fluorescence intensity of the BNQDs prepared via this method is high. Another advantage of this technique is that the particle size and quantum yield of the BNQDs can be varied by simply changing the microwave exposure time and temperature [3, 9]. Fan et al. fabricated BNQDs of uniform size via a microwave-assisted technique by rupturing boron nitride nanoplates (BNNPs) [39]. First, the synthesis of BNNPs was carried out, which involved dispersing bulk h-BN in DMF, which was ultrasonicated properly, followed by centrifugation to remove residual h-BN particles of larger size. The BNNP dispersion was subsequently exposed to 500 W microwave irradiation for 10 min, after which it was centrifuged and filtered to collect the bright yellow-colored filtrate comprising the BNQDs. The synthesized BNQDs had diameters ranging from 1.98–7.05 nm, a quantum yield of ~23.44%, and highly intense blue fluorescence, which can be attributed to their homogenous dispersion and uniform size. These BNQDs also showed excellent microwave absorption quality in addition to their remarkable optical and electrical properties. They displayed a broad effective absorption bandwidth of 5.02 GHz in the frequency region of 2—18 GHz in addition to the highest reflection loss of −19.6 dB at a frequency of 8.9 GHz. Hence, BNQDs have potential for use in the field of microwave absorption owing to their greater quantum confinement effects as well as their exceptional physical properties [39]. Jung et al. prepared hydroxyl group-functionalized BNQDs via a microwave-assisted synthesis route, and these BNQDs were utilized in bioimaging applications [40]. The synthesis of BNQDs was carried out following a defect engineering technique that exploits the impingement process of heating Fe NPs followed by exposure to microwave irradiation. Furthermore, edge hydroxyl functionalization was carried out to increase the intracellular absorption of the prepared BNQDs for bioimaging in cells. The processing of Fe NPs was strategically selected to generate defects in h-BN during the synthesis procedure. To create defects in the h-BN layers, h-BN was mixed with Fe_3O_4 and acetonitrile, and then the reaction mixture was heated to a temperature of 1000 °C in a nitrogen atmosphere. Furthermore, the heated Fe NPs on the h-BN layers created defects, which were spread by ultrasonication and microwave irradiation, leading to the formation of BNQDs. The Fe NPs that induced defects were promptly and effortlessly removed with the help of an external permanent magnet. The as-synthesized BNQDs were then treated with water vapor to functionalize the edges with hydroxyl groups in a furnace at 900 °C for 30 min. This process ultimately yields hydroxyl-functionalized BNQDs, which have low cytotoxicity, exceptional photoluminescence, and remarkable cell viability [40]. Liu et al. also employed a microwave-assisted hydrothermal synthesis route to produce sulfur-mediated BNQDs utilizing boric acid, thiourea, l-cysteine, and melamine as precursors [41]. l-Cysteine-S-BNQDs were prepared by dissolving l-cysteine, boric acid, and melamine in distilled water followed by microwave irradiation (1000 Hz) at 120 °C for 70 min. The resulting mixture was centrifuged and repeatedly rinsed to remove any remaining precursors, and finally, the produced BNQDs were collected. Thiourea-S-BNQDs were also prepared via the same procedure, except thiourea was used in place of l-cysteine. HRTEM analysis was used to study the morphologies of these two different types of S-BNQDs. The S-BNQDs prepared with L-cysteine as the precursor had an average diameter of 9.2 nm with uniform lattice fringes of 0.72 nm, whereas the S-BNQDs prepared with the thiourea precursor had an average diameter of 9.8 nm and lattice fringes of 0.76 nm [41]. On the basis of the above discussion, it can be concluded that the microwave-assisted synthesis method is an effective technique for producing BNQDs, as it ensures high quality and uniform distribution.

3 Properties

3.1 Photoluminescence Properties

The remarkable photoluminescence exhibited by BNQDs arises from their quantum confinement effect, resulting in an enlarged energy gap between the conduction band (CB) and the valence band (VB). Two primary factors influence the photoluminescence performance of BNQDs: the particle size and the defects present in the synthesized BNQDs [3, 9]. For example, Liu et al. synthesized BNQDs with average lateral sizes of approximately 2.0 ± 0.2, 2.8 ± 0.3, and 4.1 ± 0.2 nm by using three different organic solvents with different polarities, namely, NMP (10.2), DMF (6.4), and ethanol (4.3), respectively [42]. The authors reported that the BNQDs synthesized using ethanol and DMF as solvents (size: 4.1 ± 0.2 and 2.8 ± 0.3 nm) display intense blue fluorescence under irradiation with violet light, whereas the BNQDs prepared using NMP as the solvent (size: 2.0 ± 0.2 nm) displayed green fluorescence of comparatively high intensity [42]. This difference in color and fluorescence intensity of the BNQDs is attributed to their size difference resulting from the variation in the polarities of the solvent used. The high polarity of the solvent used enhances the h-BNNS and organic solvent molecule interactions, subsequently leading to the formation of smaller BNQDs. The small particle size of the BNQDs eventually shifts the wavelength (generally redshifts) in their fluorescence emission spectra. Moreover, the shift in the photoluminescence of the synthesized BNQDs is also correlated with the structural defects or oxygen present in them [42]. According to the substantive findings in the literature, with increasing solvent polarity, the number of defects present in the prepared BNQDs also increases, which leads to a higher emission wavelength. Furthermore, the defects induced as a result of carbon replacing nitrogen vacancies (3-B and 1-B centers) and the carbene assembly present at zigzag edges [9] also affect the photoluminescence properties of BNQDs. It is expected that increasing the pH of the solvents used to increase alkalinity will increase the occurrence of zigzag edges, and the contribution of 1, 3 B centers will decrease; on the other hand, reducing the pH of the solvents will have the opposite effect, subsequently increasing the photoluminescence of the synthesized BNQDs [24, 43–45]. Li et al. additionally explored the contribution of the zigzag structures and/or from the 1, 3 B centers on photoluminescence as the pH varied from 5 to 11 [16]. When the pH varied from 5 to 11, the photoluminescence peak of the BNQDs shifted from 396.8 to 401.1 nm, and the peak intensity decreased to 29%. However, the photoluminescence peak shifted back to 395.0 nm, and an additional decrease (8.4%) in the peak intensity was observed when the pH decreased from 11 to 5 [16]. Studies on the optical properties of BNQDs are at the initial stage in the current scenario, and the emission fluorescence wavelength is in the spectral range of 400—550 nm (mostly in the blue and green regions). Hence, the photoluminescence of the as-explored BNQDs at the current time is relatively distinct from that of graphene quantum dots (GQDs), which display a wide range of emission wavelengths varying from violet to red fluorescence. In the future, controlling the photoluminescence of synthesized BNQDs could be more promising and effective. For example, BNQDs can be modified by organic molecules or BNQDs doped with other elements, such as chloride, fluoride, phosphorus, and sulfur, to modulate the photoluminescence spectra of BNQDs with respect to GQDs, which will subsequently broaden the applications of BNQDs in various fields [16, 46].

3.2 *Biocompatibility and Cytotoxicity*

Biocompatibility and minimal toxicity to living cells are essential prerequisites for the use of nanomaterials in biomedical applications. Notably, boron nitride quantum dots (BNQDs), composed primarily of boron (B) and nitrogen (N), exhibit superior eco-friendliness and reduced toxicity compared with conventional heavy metal quantum dots (QDs), such as zinc selenide (ZnSe), cadmium selenide (CdSe), and cadmium telluride (CdTe) QDs. This inherent property renders BNQDs suitable for both living organisms and the environment. Hence, they provide diverse potential candidates in the field of biomedical applications, such as genetic detection, cell imaging, drug delivery, biosensing, and diagnosis [3, 9]. Many researchers have reported studies on the biocompatibility and cytotoxicity of BNQDs. Allwood et al. studied the cytotoxicity of BNQDs by using widely used mammalian cells, i.e., MDCKII cells, and demonstrated that 0–40 μg/mL BNQDs have extremely low biotoxicity to these cells [24]. Liu et al. examined the cytotoxicity of BNQDs by using 3-(4,5-dimethylthiazol-2-yl)-2,5-diphenyl tetrazolium bromide (MTT) [42]. The survival rates of He-La cells injected with various concentrations (0–350 μg mL^{-1}) of BNQDs were compared with those of untreated He-La cells. The survival rate of the He-La cells was ~93% when the concentration of BNQDs was 350 μg mL^{-1}, which indicated their low cytotoxicity and decent biocompatibility [42]. Furthermore, Liu et al. treated He-La cells with 12.5, 25, 50, and 100 μg mL^{-1} BNQDs for 24, 48, and 72 h, respectively, and calculated the survival rates of the cells via the MTT assay [23]. The results revealed that as the concentration of BNQDs increased, their cytotoxicity towards He-La cells marginally increased, and at a concentration of 100 μg mL^{-1}, the survival rate of the cells remained above 75% even after treatment for 72 h [23]. Thus, BNQDs have very low toxicity and decent biocompatibility with living cells, which proves that BNQDs are favorable components for use in biomedical applications. The MDCKII cells were also utilized by Lin et al. to evaluate the cytotoxicity of BNQDs [24]. They reported that BNQDs were not harmful to MDCKII cells, demonstrating their low cytotoxicity. Even after extended exposure to a low dose of BNQDs, no apparent effect was observed, and the potential candidacy of these BNQDs in bioimaging applications was demonstrated via luminous cell imaging. The MDCKII cells were cultured with BNQDs for 24 h before confocal imaging, and the images of the MDCKII cells stained with these BNQDs implied that they are even suitable for high-contrast bioimaging applications [24]. The optical properties of these BNQDs (at 280–420 nm) were studied via photoluminescence (PL) studies. The excitation-induced photoluminescence behavior of the emission peaks of the BNQDs was similar to that of the BNQDs described in the previous section. These BNQDs showed better photostability than conventional organic dyes and other semiconductor quantum dots did, as demonstrated by the lack of significant change in fluorescence intensity. Intense blue fluorescence was observed in the cytoplasm of the cells, which is indicative of the internalization of the BNQDs by the He-La cells, indicating that they are prominent bioimaging probes [9, 23, 24, 47, 48].

3.3 Electrochemical Luminescence Properties

Electrochemical luminescence (ECL) is an inherent property of BNQDs, allowing them to transform electrical energy into radiant energy. In the ECL process, an electroactive species undergoes conversion, emitting a fluorescent signal when an appropriate voltage is applied to the electrode [3, 9]. Ding et al. first reported the ECL properties of Si QDs in 2002, which opened a new field of research to discover the properties of other QDs [49]. BNQDs are considered to be very promising electroluminescent materials owing to their exceptional chemical stability and decent biocompatibility. In the past few years, various researchers have studied the ECL properties of BNQDs and regulated and improved their ECL performance by introducing several modifications and doping methods. Liu et al. studied and reported the ECL behavior of BNQDs and examined the influence of the presence of coreactants, including H_2O_2, L-cysteine, S_2O_8, citric acid, and glutathione, pH, and applied voltage on the ECL intensities of synthesized BNQDs [42]. The results revealed that L-cysteine is the most operative coreactant for the electrochemical luminescence of BNQDs, and when the L-cysteine amount was 5 mL, the pH of the buffer was 7, the operating cyclic voltammetry (CV) voltage was between (−2.0)-0 V, and the BNQD/L-cysteine system showed the strongest ECL intensity [42]. This observation implied that the stability and strength of the ECL signals of BNQDs can be tuned to use them extensively in the fields of sensing and imaging [41, 42, 50]. Liu et al. also used thiourea and L-cysteine as distinct sulfur precursors to fine-tune the band gap and nanostructure of synthesized BNQDs, thereby enhancing their photoelectric performance. These modified BNQDs exhibit robust and stable electrochemiluminescence (ECL) signals, broadening the potential applications of BNQDs. Although the ECL represents another remarkable property of BNQDs, research in this area is still in its early stages. Precise control and modulation of the ECL of BNQDs remain challenging, presenting an exciting avenue for further investigation. BNQDs exhibit a large band gap (5.0–6.0 eV) compared with that of other QDs, which opens another scope of investigation to decrease or moderate their large band gap via different techniques, such as doping with heteroatoms, edge configurations, and chemical functionalization, which may subsequently increase their ECL performance [3, 41, 42, 50].

4 Biomedical Applications

4.1 Genetic Detection

The genetic detection technique has played a progressively noteworthy role in primary diagnosis, effectiveness assessment, and postoperative surveillance of some diseases; hence, over the past few years, it has emerged as a focal point of interest within the modern medical field. Liu et al. fabricated an enhanced chemiluminescence (ECL) sensing system utilizing sulfur-doped BNQDs (S-BNQDs), which relies on surface plasmon-coupled enhanced chemiluminescence (SPC-ECL) [41]. This system was exploited for the identification of BRAF genes. In this system, a target deoxyribonucleic acid (DNA) interacts with hairpin DNAs (H1 and H2). Upon capture of this pair of hairpin DNAs along with BRAF genes, multiple H1–H2 duplexes become immobilized on the electrode surface after several cycles. To enhance the signal, gold nanoparticles (AuNPs) are labelled with H2 DNA, forming a signal probe that adsorbs onto the surface of S-BNQDs. The resulting resonance between the AuNPs and S-BNQDs intensifies the QD-ECL signal. Owing to the

SPC-ECL effect, the signal intensity of the ECL of the prepared S-BNQDs remarkably improved as the target DNA concentration increased. Furthermore, the effect of the BRAF gene concentration on the ECL intensity was also studied, and the results revealed that the fabricated DNA-sensing system could enumerate the BRAF gene in the range of 1 pmol/L to 1.5 nmol/L, with a minimum detection limit of 0.3 pmol/L. [41] This type of ECL-based sensor prepared with S-BNQDs has prominent scopes in the medical field. Qin et al. used BNQDs as a coreagent with a nanocomposite (rGO@Au@RueSiO_2) to increase the ECL signal intensity and fabricated an enhanced chemiluminescence (ECL)-based immunosensor from the rGO@Au@RueSiO2/BNQD heterostructure, which was employed for the detection of alpha-fetoprotein (AFP) [51]. The fabricated ECL sensor demonstrated high stability, exceptional selectivity, notable specificity, and prominent reproducibility for detecting AFP antigens. Thus, this reported ECL sensor is an efficient way to detect actual APF antigens [51]. Liu et al. developed an electrochemiluminescence (ECL) biosensor utilizing boron nitride quantum dots (BNQDs) for detecting the Shiga toxin-producing *Escherichia coli* (STEC) gene in human blood [52]. The sensor capitalizes on the ECL resonance energy transfer (ECL-RET) mechanism of synthesized orthorhombic BNQDs, coupled with the surface plasmon coupling (SPC) effect of gold nanoparticles (AuNPs). In this setup, BNQDs serve as ECL luminophores, whereas AuNPs act as luminescence quenchers because of energy transfer from BNQDs to AuNPs. Upon hybridization with the target DNA, the hairpin DNA loop attached to the BNQDs/AuNPs opens, resulting in a linear configuration and increased spacing between the BNQDs and AuNPs. Consequently, ECL-RET is replaced by the SPC-ECL effect, leading to nearly sixfold amplification of the ECL signal. This innovative ECL-RET/SPC-ECL sensing mode significantly enhances the accuracy and sensitivity of the BNQD-based ECL sensor. The fabricated biosensor can accurately measure the STEC gene in the range of 1 pmol L^{-1} to 5 nmol L^{-1} [52].

4.2 Biosensing

Kong et al. successfully derived BNQDs from boric acid and amino acids, and the latter was used as a nitrogen precursor by employing a hydrothermal synthesis route [53]. They used synthesized BNQDs as on-off-on fluorescent sensors for sensing ascorbic acid (AA) in human serum and in real water samples by employing Fe(III) as the medium. The quenching strategy using BNQDs termed "turn-off" by Fe(III) was due to the inner filter effect (IFE). IFE emerges because of the spectral overlap between the emission spectrum of the fluorophore species and the absorbance spectrum of the quencher species. After ascorbic acid was added to the system, the fluorescence recovered because of a redox reaction between Fe(III) and AA, which decomposes the BNQD-Fe(III) complex. This restoration of fluorescence was termed the "turn-on" switch of the fabricated sensor. The sensor exhibited a detection limit at a concentration of 0.0833 μM for AA, which proves that it is highly efficient and sensitive [53]. Han et al. reported the application of fluorescent probes using 2D-BNQDs for the estimation of an important biomarker, alkaline phosphatase enzyme (ALP) [54]. The sensing procedure included decreasing the intensity of fluorescence using synthesized BNQDs after the addition of Fe(III) ions (turn-off strategy) and subsequently restoring the intensity of fluorescence (turn-on strategy) with the incorporation of pyrophosphate and ALP. The fluorescence restoration of BNQDs results from the production of orthophosphate, which causes breakdown of the BNQD-Fe(III) complex [54]. Peng et al. applied the MnO_2 nanosheet (NS) IFE on BNQDs synthesized via a hydrothermal technique to detect glutathione (GSH) [55]. Glutathione (GSH), a nonprotein biothiol that is abundant in eukaryotic cells, plays pivotal roles in various biological functions. These functions encompass cell growth regulation, maintenance of the cellular antioxidant defense system, and modulation of protein function. The IFE comprises the excitation or absorption of the emitted beam present in this sensing system, which occurs due to the overlay of the absorption spectrum of GSH with the fluorescence spectrum of the BNQDs. The fluorescence intensity does not vary in the presence of only GSH, indicating the occurrence of very few interactions among GSH and BNQDs and indicating the remarkable photostability of the BNQDs. However, MnO_2 may decompose to form Mn^{2+} due to the addition of GSH, which subsequently deteriorates the IFE, resulting in the gradual recovery of fluorescence. This is affected by the amount of GSH present in the system. On the basis of the IFE, a fluorescent sensor was fabricated with BNQDs, which can detect GSH promptly within the concentration range of 0.5–250 μM with considerable sensitivity [55]. Yang et al. fabricated an efficient BNQD-Eu(III) sensor for tetracycline (TC) detection, as this antibiotic is extensively used for the treatment of microorganism infection, but its overeating causes the presence of residues in the human body, which eventually affects liver function and leads to bacterial resistance [56]. The intense blue fluorescence of the BNQDs was quenched, and the red fluorescence of Eu(III) intensified when TCs were introduced into the hybrid sensor probe of the BNQDs and Eu(III) as a consequence of the IFE and photoinduced electron transfer [56]. Zhan et al. designed gold nanoparticles (AuNPs) on BNQDs as a fluorescent sensor, which was employed for the sensing of acetylcholinesterase (AChE) activity [57]. The plausible mechanism for AChE detection was based on the fluorescence resonance energy transfer (FRET) phenomenon occurring between the BNQDs and AuNPs. In this process, acetylthiocholine (ATCh) is decomposed by acetylcholinesterase to yield thiocholine (TCh), which subsequently reduces chloroauric acid ($HAuCl_4$) to finally yield AuNPs, and in situ formation of TCh-BNQDs/AuNPs occurs, causing fluorescence quenching of the BNQDs. Furthermore, an organophosphorus insecticide, paraoxon, was introduced to decrease AChE activity and hinder the enzymatic hydrolysis reaction, resulting in fluorescence quenching of the BNQDs. The UV–Vis absorption spectrum of the in situ-produced AuNPs and the fluorescence spectrum of ATCh-BNQDs overlapped to a large extent, which caused a decreasing trend in intensity when BNQDs were used. Compared with ATCh-BNQDs/HAuCl4, the solution of BNQDs/ATCh/$HAuCl_4$ also strongly quenched AChE. This difference in quenching efficiency arises from the presence of the unreacted –SH termini in the

ATCh-BNQDs adjacent to neighboring AuNPs, resulting in TCh-BNQD/AuNP assembly via intermolecular cross-linking with the Au–S bond, which additionally quenches the fluorescence [57]. This fabricated fluorescent sensor was established to have diverse applications in various fields, such as inhibitor screening, clinical diagnosis, environmental toxic pollutant detection, and food safety analysis.

The basic principle of ECL involves the emission of a detectable signal from a marker in response to an applied voltage, as discussed in the previous section. The ECL technique possesses many unique features, such as the absence of any radioisotopes, enormously low detection limits for the markers, optical excitation eluding the environment from scattered source light or sample autofluorescence, and the requirement of very little time. Because of these distinct features, ECL is very efficient and beneficial compared with other analytical measurement techniques. BNQDs are widely used as ECL sensors for the sensitive and highly efficient detection of enzymes, genes, ascorbic acid, uric acid, dopamine, etc. [9] Wang et al. designed a BNQD-supported ECL biosensor, which was employed for the estimation of a legume protein called concanavalin A (Con A) [58]. The interaction of Con A with carbohydrates plays a crucial role in the identification of cancer cells; hence, an effective estimation of the concentration of Con A is highly important for clinical diagnosis. In this fabricated ECL sensor, BNQDs were used as a coreagent in a poly luminol-reduced graphene oxide system (PLA-rGO) to increase the ECL intensity of luminol. A remarkable tenfold increase in the ECL intensity was achieved compared with that of the individual PLA-rGO, which indicated the important role of these BNQDs as coreagents. After the immobilization of Glu/PLA-rGO, bovine serum albumin (BSA) was introduced to the electrode to eliminate the nonspecific binding effects. Eventually, the ECL intensity of luminol decreased as the concentration of Con A on the electrode was varied by means of a specific carbohydrate-Con A interaction, which was utilized for Con A estimation [58]. In basic media, the oxidation of luminol occurs to produce luminol$^{\bullet-}$ and BNQDs undergo electrochemical oxidation to give excess BNQDs-NH$^{+\bullet}$. The produced BNQDs-NH$^{+\bullet}$ further undergo reduction in the basic medium to yield BNQDs-N$^{\bullet}$ which interacts with luminol$^{\bullet-}$ to produce an intermediate excited species (AP^{2-*}) [9, 42]. This excited state species emits light while recurring to the ground state. Con A interferes with these reaction steps, which decreases the ECL intensity, verifying the presence of Con A in the system. The fabricated biosensor exhibited high sensitivity and a very low detection limit (0.15 pg.mL^{-1}) [58]. Xing et al. used BNQDs as coreagents to increase the ECL intensity of ruthenium(II) tris(2,2′-bipyridyl), denoted by $[Ru(bpy)_3]^{2+}$ [59]. The resulting $[Ru(bpy)_3]^{2+}$/BNQD system was employed for the detection and estimation of dopamine (DA), which acts as a neurotransmitter and plays a crucial role in human physiology. In the presence of DA, the intensity of the optical signal decreases, which is due to the interaction of oxidized $[Ru(bpy)_3]^{2+}$ and deprotonated BNQDs. ECL quenching occurs as a consequence of the formation of o-benzoquinone from the oxidation of DA, which interacts with the excited state of $[Ru(bpy)_3]^{2+}$ [59].

4.3 Fluorescence Bioimaging

BNQDs possess analogous physical and chemical properties to those of conventional QDs; however, they also have distinctive benefits, such as stable fluorescence properties, low cytotoxicity, and decent biocompatibility. This makes them an advanced replacement in place of conventional QDs, which have been prominent materials in the field of biological research. Lei et al. successfully synthesized BNQDs in 2015 and used them to detect He-La cells [48]. First, the He-La cells were cultured in DMEM or high-glucose medium along with the synthesized BNQDs at 37 °C for approximately 7 h. Bright, intense fluorescence of He-La cells was detected under a laser confocal fluorescence (LCF) microscope. In addition, the experimental data also showed decent overlay of the fluorescence of LysoTracker (lysosome red fluorescent probe) with the BNQD probe in He-La cells, indicating the entry of BNQDs into the cell body via endocytosis, along with the admission of these BNQDs to the cytoplasm via the cell membrane instead of the nuclei of the cells. Therefore, these BNQDs were utilized in cell marking because they do not interrupt the destruction of gene sequences, which further verified the low cytotoxicity of the synthesized BNQDs [48]. Similarly, Lin et al. marked MDCKII cells with BNQDs and 4,6-diamidino-2-phenylic-dole (DAPI) to demonstrate the potential utilization of BNQDs in high-contrast bioimaging applications [24]. Thangasamy et al. carried out the synthesis of BNQDs from layered bulk h-BN by employing a supercritical fluid processing technique, which resulted in an average lateral size of 5 nm and blue luminescence [21]. A homogeneous reaction mixture, termed a supercritical fluid (SCF), was formed by increasing the temperature and pressure of the solvent used beyond its critical values. The strong blue luminescent property of the synthesized BNQDs was used for the detection of gram-negative bacterial cells. The authors first strained the collected stationary phase cells with BNQDs (~100 μg/mL) for approximately 15 min, followed by epifluorescence microscopy imaging, which revealed that the gram-negative cells were selectively marked by BNQDs, whereas the gram-positive cells were not strained by these BNQDs. The coating of the BNQDs on the bacterial surface was also verified via FESEM analysis and confocal microscopy [21]. The bacterial cells possess a cell wall outside the lipid bilayer and peptidoglycan layer, which differs from living animal cells. Nonetheless, gram-negative cells are composed of a lipid layer outside the peptidoglycan layer, which is absent in the case of gram-positive cells. Therefore, the selectivity of the synthesized BNQDs towards gram-negative cells can be attributed to the presence of this outer lipid layer [21]. Furthermore, they also repeated the same experiment with gram-negative and gram-positive bacterial cells by staining them with graphene quantum dots (GQDs) and compared the performance with that of BNQDs. Interestingly, the performance of the synthesized BNQDs was superior to that of the GQDs, indicating the effectiveness and suitability of the synthesized BNQDs for the selective detection of gram-negative cells [21]. Yang et al. explored emerging metal-doped boron quantum dots (Co@BQDs) for the detection of lactate FL and fluorescence bioimaging [60]. The Co atom is attached to the B atom via an effective Co–B bond that overpowers the B–O reaction, thereby protecting the electronic structures of delaminated 2D boron from being oxidized and decaying after interaction with oxygen. The doping of BQDs with Co atoms also increased their structural stability and electroactivity. The synthesized Co@BQDs exhibited greater colloidal and fluorescent stability than did the individual BQDs owing to the formation of Co–B bonds and the limitation of the B–O reaction. The Co@BQDs displayed intense and stable fluorescence that can be observed with the naked eye and were utilized for visual fluorescence imaging of phosphate-buffered saline (PBS) dispersions. The fluorescence imaging tests were carried out by evaluating the visual fluorescence colors of the Co@BQD-drenched living glass fishes, fresh willow buds, and F. velutipes. A feasible fluorescent bioprobe of lactate was fabricated with Co@BQDs by means of enzymatic and cascade oxidation, which demonstrated selective and profound fluorescence responses with the addition of lactate, facilitating the effective detection of lactate in biosamples via fluorescence spectral analysis [60]. Hence, this bioprobe has the potential for durable tracking and bioimaging in living cells in imminent studies.

4.4 Photothermal Therapy

Photothermal therapy (PT), in which photothermal species are employed to absorb near-infrared (NIR) radiation to generate photothermal effects that eventually abolish targeted cancer cells, is an evolving method for the treatment of cancerous tumors [61]. This treatment method has great potential in tumor therapy because of its effectiveness and high precision, short span of treatment time, nonintrusiveness, low side effects, etc. In addition to destroying cancerous cells, the photothermal outcome can also produce sound waves that are detectable and can be transformed into imaging signals, and the process is termed photoacoustic imaging. This process benefits from the high contrast of optical imaging and extreme spatial resolution of ultrasound imaging, subsequently providing an advanced nonintrusive and nonionizing imaging technique for the diagnosis of tumors [61–64]. Hence, the design of innovative photothermal agents that offer tremendous photothermal effects and decent biocompatibility is highly important. Guo et al. designed PEG-modified BQDs (BQDs-PEG) and explored their applications in photothermal therapy as well as the photoacoustic imaging technique [61]. The synthesized BQDs-PEG can effectively generate photothermal effects under NIR irradiation and can remarkably destroy cancerous cells and obstruct the growth of tumors via photothermal effects, which was verified by both in vitro and in vivo experiments. The in vitro photothermal properties of BQDs-PEG were studied by irradiating various concentrations of BQDs-PEG with a laser comprising a power density of 1 W cm^{-2} for approximately 10 min and monitoring the temperature. The temperature of the BQD-PEG solution increased with increasing solution concentration, and the temperature of the solution increased to 52 °C within 10 min of irradiation at a concentration of 100 μg mL^{-1}. However, the temperature of only water did not notably increase upon exposure to the same irradiation for 10 min. The calculated PT conversion efficiency of these BQD-PEGs was ~23.7%. In addition, the BQDs-PEG showed insignificant changes in their photothermal conversion performance even after four cycles of laser irradiation, which implied their high photothermal stability and demonstrated their ability to convert light into heat energy efficiently. The efficacy of BQD-PEG in PT cell therapy was assessed via the standard MTT method. The BQDs-PEG had no noteworthy harmful effect on cell viability in the absence of laser irradiation, indicating their good biocompatibility and low cytotoxicity. In contrast, when 4T1 cancer cells were exposed to 808 nm laser radiation, the cancer cell viability (CCV) of these cancer cells decreased excessively. The 4T1 CCV also decreased as the concentration of BQDs-PEG increased, which verified their photothermal effect and efficacy in destroying tumor cells. The 4T1 cells were stained with calcein-AM and propidium iodide to visualize the cell viability optically. In the absence of laser irradiation, the cells displayed green fluorescence, indicating the biocompatibility of the BQDs-PEG. On the other hand, a large quantity of red fluorescence was observed when the cells were exposed to 808 nm laser irradiation and treated with BQDs-PEG [61]. This observation indicated that the majority of the cancerous cells were killed during the treatment. The outstanding photothermal effect of the synthesized BQDs-PEG encouraged the authors to study the PA imaging efficiency of these BQDs-PEGs. Aqueous solutions of BQDs-PEG at several concentrations were prepared for the measurement of in vitro photoacoustic (PA) signals. The intensity of the PA signals increased as the concentration of the BQD-PEG solution used increased. Because of the strong in vitro PA imaging effects of BQDs-PEG, they were further utilized as PA contrast agents in animals. The BQDs-PEG were injected into nude mice comprising 4T1 tumors, and at different durations, photoacoustic images of the tumor region were collected. In the tumor region, the photoacoustic signals significantly increased following the administration of BQDs-PEG to the mice. This enhancement suggests the effective accumulation of BQDs-PEG at the tumor site, likely facilitated by superior penetration and retention mechanisms. These results establish the good potential of BQD-PEG as a contrast agent for photoacoustic imaging. The in vivo PT effect of the synthesized BQDs-PEG was also studied by using 4T1 tumor-bearing nude mice as animal prototypes. The hemolysis tests of BQDs-PEG revealed that the hemolysis rate was <1.8%, although the concentration of the solution was very high (~200 μg mL^{-1}), indicating that they do not show hemolysis and exhibit excellent blood compatibility [61]. Thus, the BQD-PEG system has promising potential for use in photothermal cancer treatment. Consoli et al. synthesized water-dispersible BNQDs comprising phenolic and carboxylic groups [65]. Carbon-BNQDs are synthesized via a single-step solvent-assisted method from boronic acid and urea. The obtained carbon-BNQDs exhibit high photothermal conversion efficiency and pH-responsive luminescence. The biocompatibility of the carbon-BNQDs was examined by using colorectal adenocarcinoma cells (Caco-2), which showed protective effects against chemical and oxidative stress. The cell protective effects of the carbon-BNQDs were also evaluated via analysis of the amplicon mRNA sequence for the quantification of 56 genes involved in oxidative stress [65]. Wang et al. synthesized BQDs via a liquid-phase exfoliation technique and then successfully fabricated a BQD-based all-optical modulator. The obtained BQDs were first deposited on the surface of a microfiber, and this derived BQD-microfiber assembly displayed a proficient photothermal conversion rate along with a very high photothermal response [66].

5 Conclusion and Outlook

The present discussion provides a brief summary of boron nitride quantum dots, including the available synthesis techniques, studied properties and their emerging applications in the biomedical field. In addition to their distinctive and advantageous properties, these QDs also exhibit greater dispersibility, making them adequate for solution-based procedures. Herein, we explored the various synthesis techniques used by different groups of researchers to achieve BNQDs, the advantages, and the limitations of these techniques. Some unique properties of BNQDs, including photoluminescence, electrochemical luminescence, biocompatibility, and cytotoxicity, which are prerequisites for their applications in the biomedical field, are discussed, with specifics to obstacles regarding the adjustment of photoluminescence, ECL intensities, etc., also included. An in-depth analysis of the usage of BNQDs in various biomedical applications, such as genetic detection, biosensors, bioimaging, and photothermal therapy, was subsequently performed. The overview of boron quantum dot technology augments new pathways for the development of advanced materials with improved functionality and performance, making them an exciting area to focus on in terms of research in nanotechnology and materials science. In the future, controlling the photoluminescence of synthesized BNQDs could be more promising and effective. The adjustment of these BNQDs via organic molecules or mixing with agents such as chloride, fluoride, phosphorus, and sulfur aims to improve and control their structural and electrical properties, similar to those of GQDs, thereby widening the implementation of BNQDs in different fields.

Acknowledgement The authors would like to acknowledge CSIR-North East Institute of Science and Technology, Jorhat, Assam, India.

References

1. Ahmad, P., Khandaker, M. U., Muhammad, N., Khan, G., Rehman, F., Khan, A. S., Ullah, Z., Khan, A., Ali, H., Ahmed, S. M., Rauf Khan, M. A., Iqbal, J., Khan, A. A., & Irshad, M. I. (2019). Fabrication of hexagonal boron nitride quantum dots via a facile bottom-up technique. *Ceramics International, 45*(17, Part B), 22765–22768.
2. Chen, F., Gao, W., Qiu, X., Zhang, H., Liu, L., Liao, P., Fu, W., & Luo, Y. (2017). Graphene quantum dots in biomedical applications: Recent advances and future challenges. *Frontiers in Laboratory Medicine, 1*(4), 192–199.

3. Zhang, X., An, L., Bai, C., Chen, L., & Yu, Y. (2021). Hexagonal boron nitride quantum dots: Properties, preparation and applications. *Materials Today, Chemistry, 20*, 100425.
4. Ekimov, A. I. (1981). Quantum size effect in three-dimensional microscopic semiconductor crystals. *JETP Letters, 34*, 345.
5. Gupta, G. H., Kadakia, S., Agiwal, D., Keshari, T., & Kumar, S. (2024). Borophene nanomaterials: Synthesis and applications in biosensors. *Materials Advances, 5*(5), 1803–1816.
6. Ogitsu, T., Schwegler, E., & Galli, G. (2013). β-Rhombohedral boron: At the crossroads of the chemistry of boron and the physics of frustration. *Chemical Reviews, 113*(5), 3425–3449.
7. Hao, J., Tai, G., Zhou, J., Wang, R., Hou, C., & Guo, W. (2020). Crystalline semiconductor boron quantum dots. *ACS Applied Materials & Interfaces, 12*(15), 17669–17675.
8. Lee, J. W., Kshetri, T., Park, K. R., Kim, N. H., Park, O.-K., & Lee, J. H. (2021). Covalently bonded boron nitride quantum dot and reduced graphene oxide composite electrode for highly efficient supercapacitors. *Composites Part B: Engineering, 222*, 109089.
9. Rawat, J., Sajwan, D., Garimella, S. V., Sharma, H., & Dwivedi, C. (2023). Boron Nitride quantum dots: A rising star in sensing applications. *Nano Trends, 2*, 100008.
10. Chen, L., Zhang, X., Zhao, Z., Wang, F., Huang, Y., Bai, C., An, L., & Yu, Y. (2021). Controllable preparation of boron nitride quantum dots with small size and strong blue photoluminescence. *Colloids and Surfaces A: Physicochemical and Engineering Aspects, 614*, 126181.
11. Angizi, S., Shayeganfar, F., Azar, M. H., & Simchi, A. (2020). Surface/edge functionalized boron nitride quantum dots: Spectroscopic fingerprint of bandgap modification by chemical functionalization. *Ceramics International, 46*(1), 978–985.
12. Umrao, S., Maurya, A. K., Shukla, V., Grigoriev, A., Ahuja, R., Vinayak, M., Srivastava, R. R., Saxena, P. S., Oh, I. K., & Srivastava, A. (2019). Anticarcinogenic activity of blue fluorescent hexagonal boron nitride quantum dots: As an effective enhancer for DNA cleavage activity of anticancer drug doxorubicin. *Materials Today, Bio, 1*, 100001.
13. Angizi, S., Hatamie, A., Ghanbari, H., & Simchi, A. (2018). Mechanochemical green synthesis of exfoliated edge-functionalized boron nitride quantum dots: Application to vitamin C sensing through hybridization with gold electrodes. *ACS Applied Materials & Interfaces, 10*(34), 28819–28827.
14. Dehghani, A., Madadi Ardekani, S., Lesani, P., Hassan, M., & Gomes, V. G. (2018). Two-photon active boron nitride quantum dots for multiplexed imaging, intracellular ferric ion biosensing, and pH tracking in living cells. *ACS Applied Bio Materials, 1*(4), 975–984.
15. Alzakia, F. I., & Tan, S. C. (2021). Liquid-exfoliated 2D materials for optoelectronic applications. *Advanced Science, 8*(11), 2003864.
16. Li, H., Tay, R. Y., Tsang, S. H., Zhen, X., & Teo, E. H. T. (2015). Controllable synthesis of highly luminescent boron nitride quantum dots. *Small, 11*(48), 6491–6499.
17. Kumar, R., Singh, R. K., Yadav, S. K., Savu, R., & Moshkalev, S. A. (2016). Mechanical pressure induced chemical cutting of boron nitride sheets into boron nitride quantum dots and optical properties. *Journal of Alloys and Compounds, 683*, 38–45.
18. Hod, O. (2012). Graphite and hexagonal boron-nitride have the same interlayer distance. Why? *Journal of Chemical Theory and Computation, 8*(4), 1360–1369.
19. Kovtyukhova, N. I., Wang, Y., Lv, R., Terrones, M., Crespi, V. H., & Mallouk, T. E. (2013). Reversible intercalation of hexagonal boron nitride with Brønsted acids. *Journal of the American Chemical Society, 135*(22), 8372–8381.
20. Štengl, V., Henych, J., & Kormunda, M. (2014). Self-assembled BN and BCN quantum dots obtained from high intensity ultrasound exfoliated nanosheets. *Science of Advanced Materials, 6*(6), 1106–1116.
21. Thangasamy, P., Santhanam, M., & Sathish, M. (2016). Supercritical fluid facilitated disintegration of hexagonal boron nitride nanosheets to quantum dots and its application in cells imaging. *ACS Applied Materials & Interfaces, 8*(29), 18647–18651.
22. Adschiri, T., Lee, Y.-W., Goto, M., & Takami, S. (2011). Green materials synthesis with supercritical water. *Green Chemistry, 13*(6), 1380–1390.

23. Liu, Q., Hu, C., & Wang, X. (2019). One-pot solvothermal synthesis of water-soluble boron nitride nanosheets and fluorescent boron nitride quantum dots. *Materials Letters, 234*, 306–310.
24. Lin, L., Xu, Y., Zhang, S., Ross, I. M., Ong, A., & Allwood, D. A. (2013). Fabrication and luminescence of monolayered boron nitride quantum dots. *Small (Weinheim a der Bergstrasse, Germany), 10*(1), 60–65.
25. Freeman, A., & Larkindale, J. (1969). Evidence for the formation of boron nitride-alkali metal intercalation compounds. *Inorganic and Nuclear Chemistry Letters, 5*(11), 937–939.
26. Peng, D., Zhang, L., Li, F.-F., Cui, W.-R., Liang, R.-P., & Qiu, J.-D. (2018). Facile and green approach to the synthesis of boron nitride quantum dots for 2, 4, 6-trinitrophenol sensing. *ACS Applied Materials & Interfaces, 10*(8), 7315–7323.
27. Bandyopadhyay, A., Yamijala, S. S., & Pati, S. K. (2013). Tuning the electronic and optical properties of graphene and boron-nitride quantum dots by molecular charge-transfer interactions: A theoretical study. *Physical Chemistry Chemical Physics, 15*(33), 13881–13887.
28. Amendola, V., & Meneghetti, M. (2009). Laser ablation synthesis in solution and size manipulation of noble metal nanoparticles. *Physical Chemistry Chemical Physics, 11*(20), 3805–3821.
29. Nguyen, V., Yan, L., Zhao, N., Van Canh, N., Hang, N. T. N., & Le, P. H. (2021). Tuning photoluminescence of boron nitride quantum dots via surface functionalization by femtosecond laser ablation. *Journal of Molecular Structure, 1244*, 130922.
30. Xu, Y., Yan, L., Si, J., & Zhao, N. (2019). Femtosecond laser-assisted fabrication of fluorescent boron nitride quantum dots. In *2019 IEEE international conference on Manipulation, Manufacturing and Measurement on the Nanoscale (3 M-NANO)* (pp. 7–10). IEEE.
31. Tan, C., Cao, X., Wu, X.-J., He, Q., Yang, J., Zhang, X., Chen, J., Zhao, W., Han, S., & Nam, G.-H. (2017). Recent advances in ultrathin two-dimensional nanomaterials. *Chemical Reviews, 117*(9), 6225–6331.
32. Wang, Y., & Ni, Y. (2014). Molybdenum disulfide quantum dots as a photoluminescence sensing platform for 2, 4, 6-trinitrophenol detection. *Analytical Chemistry, 86*(15), 7463–7470.
33. Sert, B., Gonca, S., Ozay, Y., Harputlu, E., Ozdemir, S., Ocakoglu, K., & Dizge, N. (2021). Investigation of the antifouling properties of polyethersulfone ultrafiltration membranes by blending of boron nitride quantum dots. *Colloids and Surfaces B: Biointerfaces, 205*, 111867.
34. Huo, B., Liu, B., Chen, T., Cui, L., Xu, G., Liu, M., & Liu, J. (2017). One-step synthesis of fluorescent boron nitride quantum dots via a hydrothermal strategy using melamine as nitrogen source for the detection of ferric ions. *Langmuir, 33*(40), 10673–10678.
35. Liu, B., Yan, S., Song, Z., Liu, M., Ji, X., Yang, W., & Liu, J. (2016). One-step synthesis of boron nitride quantum dots: Simple chemistry meets delicate nanotechnology. *Chemistry–A European Journal, 22*(52), 18899–18907.
36. Kamyabi, M. A., & Moharramnezhad, M. (2022). Single-step microwave synthesis of a novel ternary nanocomposite as an efficient luminophore and boron nitride quantum dots as a new coreactant for a cathodic ECL monitoring of chlorpyrifos. *Analytical Methods, 14*(7), 750–762.
37. Chu, K., Li, X., Tian, Y., Li, Q., & Guo, Y. (2022). Boron nitride quantum dots/Ti3C2Tx-MXene Heterostructure for efficient electrocatalytic nitrogen fixation. *Energy & Environmental Materials, 5*(4), 1303–1309.
38. Wang, L., Zhang, Q., Su, P., Yu, L., Bu, Y., Yuan, C., & Wang, S. (2022). Excitation-dependent ratiometric fluorescence response to mercury ion based on single hexagonal boron nitride quantum dots. *Analytica Chimica Acta, 1236*, 340585.
39. Fan, L., Zhou, Y., He, M., Tong, Y., Zhong, X., Fang, J., & Bu, X. (2017). Facile microwave approach to controllable boron nitride quantum dots. *Journal of Materials Science, 52*(23), 13522–13532.
40. Jung, J.-H., Kotal, M., Jang, M.-H., Lee, J., Cho, Y.-H., Kim, W.-J., & Oh, I.-K. (2016). Defect engineering route to boron nitride quantum dots and edge-hydroxylated functionalization for bioimaging. *RSC Advances, 6*(77), 73939–73946.

41. Liu, Y., Wang, M., Nie, Y., Zhang, Q., & Ma, Q. (2019). Sulfur regulated boron nitride quantum dots electrochemiluminescence with amplified surface plasmon coupling strategy for BRAF gene detection. *Analytical Chemistry, 91*(9), 6250–6258.
42. Liu, M., Xu, Y., Wang, Y., Chen, X., Ji, X., Niu, F., Song, Z., & Liu, J. (2017). Boron nitride quantum dots with solvent-regulated blue/green photoluminescence and electrochemiluminescent behavior for versatile applications. *Advanced Optical Materials, 5*(3), 1600661.
43. Katzir, A., Suss, J., Zunger, A., & Halperin, A. (1975). Point defects in hexagonal boron nitride. I. EPR, thermoluminescence, and thermally stimulated-current measurements. *Physical Review B, 11*(6), 2370.
44. Zunger, A., & Katzir, A. (1975). Point defects in hexagonal boron nitride. II. Theoretical studies. *Physical Review B, 11*(6), 2378.
45. Lin, L., & Zhang, S. (2012). Creating high yield water soluble luminescent graphene quantum dots by exfoliating and disintegrating carbon nanotubes and graphite flakes. *Chemical Communications, 48*(82), 10177–10179.
46. Wang, L., Li, W., Yin, L., Liu, Y., Guo, H., Lai, J., Han, Y., Li, G., Li, M., & Zhang, J. (2020). Full-color fluorescent carbon quantum dots. *Science Advances, 6*(40), eabb6772.
47. Rafiei-Sarmazdeh, Z., Jafari, S. H., Ahmadi, S. J., & Zahedi-Dizaji, S. M. (2016). Large-scale exfoliation of hexagonal boron nitride with combined fast quenching and liquid exfoliation strategies. *Journal of Materials Science, 51*, 3162–3169.
48. Lei, Z., Xu, S., Wan, J., & Wu, P. (2015). Facile preparation and multifunctional applications of boron nitride quantum dots. *Nanoscale, 7*(45), 18902–18907.
49. Ding, Z., Quinn, B. M., Haram, S. K., Pell, L. E., Korgel, B. A., & Bard, A. J. (2002). Electrochemistry and electrogenerated chemiluminescence from silicon nanocrystal quantum dots. *Science, 296*(5571), 1293–1297.
50. Dong, Y., Tian, W., Ren, S., Dai, R., Chi, Y., & Chen, G. (2014). Graphene quantum dots/l--cysteine coreactant electrochemiluminescence system and its application in sensing lead (II) ions. *ACS Applied Materials & Interfaces, 6*(3), 1646–1651.
51. Qin, D., Jiang, X., Mo, G., Feng, J., & Deng, B. (2020). Boron nitride quantum dots as electrochemiluminescence coreactants of rGO@Au@Ru–SiO2 for label-free detection of AFP in human serum. *Electrochimica Acta, 335*, 135621.
52. Liu, Y., Chen, X., Wang, M., & Ma, Q. (2018). A visual electrochemiluminescence resonance energy transfer/surface plasmon coupled electrochemiluminescence nanosensor for Shiga toxin-producing *Escherichia coli* detection. *Green Chemistry, 20*(24), 5520–5527.
53. Kong, Y., He, Y., Zhou, J., Zhong, S., & Song, G. (2020). Amino acids as the nitrogen source to synthesize boron nitride quantum dots for fluorescence turn-off-on detection of ascorbic acid. *ChemistrySelect, 5*(13), 3828–3834.
54. Han, Y., Niu, Y., Liu, M., Niu, F., & Xu, Y. (2019). A rational strategy to develop a boron nitride quantum dot-based molecular logic gate and fluorescence assay of alkaline phosphatase activity. *Journal of Materials Chemistry B, 7*(6), 897–902.
55. Peng, C., Xing, H., Fan, X., Xue, Y., Li, J., & Wang, E. (2019). Glutathione regulated inner filter effect of MnO2 nanosheets on boron nitride quantum dots for sensitive assay. *Analytical Chemistry, 91*(9), 5762–5767.
56. Yang, K., Jia, P., Hou, J., Bu, T., Sun, X., Liu, Y., & Wang, L. (2020). Innovative dual-emitting Ratiometric fluorescence sensor for tetracyclines detection based on boron nitride quantum dots and europium ions. *ACS Sustainable Chemistry & Engineering, 8*(46), 17185–17193.
57. Zhan, Y., Yang, J., Guo, L., Luo, F., Qiu, B., Hong, G., & Lin, Z. (2019). Targets regulated formation of boron nitride quantum dots—Gold nanoparticles nanocomposites for ultrasensitive detection of acetylcholinesterase activity and its inhibitors. *Sensors and Actuators B: Chemical, 279*, 61–68.
58. Wang, C., Li, M., Wang, P., & Liu, D. (2020). An electrochemiluminescence biosensor based on boron nitride quantum dots as novel coreactant for quantitative determination of concanavalin A. *Microchimica Acta, 187*(7), 409.

59. Xing, H., Zhai, Q., Zhang, X., Li, J., & Wang, E. (2018). Boron nitride quantum dots as efficient coreactant for enhanced Electrochemiluminescence of Ruthenium(II) Tris(2,2′-bipyridyl). *Analytical Chemistry, 90*(3), 2141–2147.
60. Yang, M., Jin, H., & Gui, R. (2022). Metal-doped boron quantum dots for versatile detection of lactate and fluorescence bioimaging. *ACS Applied Materials & Interfaces, 14*(51), 56986–56997.
61. Guo, T., Tang, Q., Guo, Y., Qiu, H., Dai, J., Xing, C., Zhuang, S., & Huang, G. (2021). Boron quantum dots for photoacoustic imaging-guided photothermal therapy. *ACS Applied Materials & Interfaces, 13*(1), 306–311.
62. Manohar, S., & Gambhir, S. (2020). Clinical photoacoustic imaging. *Photoacoustics*, 19.
63. Sun, Z., Zhao, Y., Li, Z., Cui, H., Zhou, Y., Li, W., Tao, W., Zhang, H., Wang, H., & Chu, P. K. (2017). As an efficient contrast agent for in vivo photoacoustic imaging of cancer. *Small, 13*(1602896), 10.1002.
64. Zhang, J., Zhang, J., Li, W., Chen, R., Zhang, Z., Zhang, W., Tang, Y., Chen, X., Liu, G., & Lee, C.-S. (2017). Degradable hollow mesoporous silicon/carbon nanoparticles for photoacoustic imaging-guided highly effective chemo-thermal tumor therapy in vitro and in vivo. *Theranostics, 7*(12), 3007.
65. Consoli, G. M., Maugeri, L., Musso, N., Gulino, A., D'Urso, L., Bonacci, P., Buscarino, G., Forte, G., & Petralia, S. (2024). One-pot synthesis of luminescent and photothermal carbon boron-nitride quantum dots exhibiting cell damage protective effects. *Advanced Healthcare Materials*, 2303692.
66. Wang, C., Chen, Q., Chen, H., Liu, J., Song, Y., Liu, J., Li, D., Ge, Y., Gong, Y., & Zhang, Y. (2021). Boron quantum dots all-optical modulator based on efficient photothermal effect. *Opto-Electronic Advances, 4*(7), 200032-1-200032-10.

Index

S. P. Kabekkodu et al. (eds.), *Functional Quantum Dots*, Lecture Notes in Nanoscale Science and Technology 40,
https://doi.org/10.1007/978-3-031-98219-4

MIX
Papier aus verantwortungsvollen Quellen
Paper from responsible sources
FSC® C105338

If you have any concerns about our products,
you can contact us on
ProductSafety@springernature.com

In case Publisher is established outside the EU,
the EU authorized representative is:
Springer Nature Customer Service Center GmbH
Europaplatz 3, 69115 Heidelberg, Germany

Printed by Libri Plureos GmbH
in Hamburg, Germany